Exploring Science

The Exploring Science series is designed to familiarize young students with science topics taught in grades 4–9. The topics in each book are divided into knowledge and understanding sections, followed by exploration by means of simple projects or experiments. The topics are also sequenced from easiest to more complex, and should be worked through until the correct level of attainment for the age and ability of the student is reached. Carefully planned Test Yourself questions at the end of each topic allow the student to gain a sense of achievement on mastering the subject.

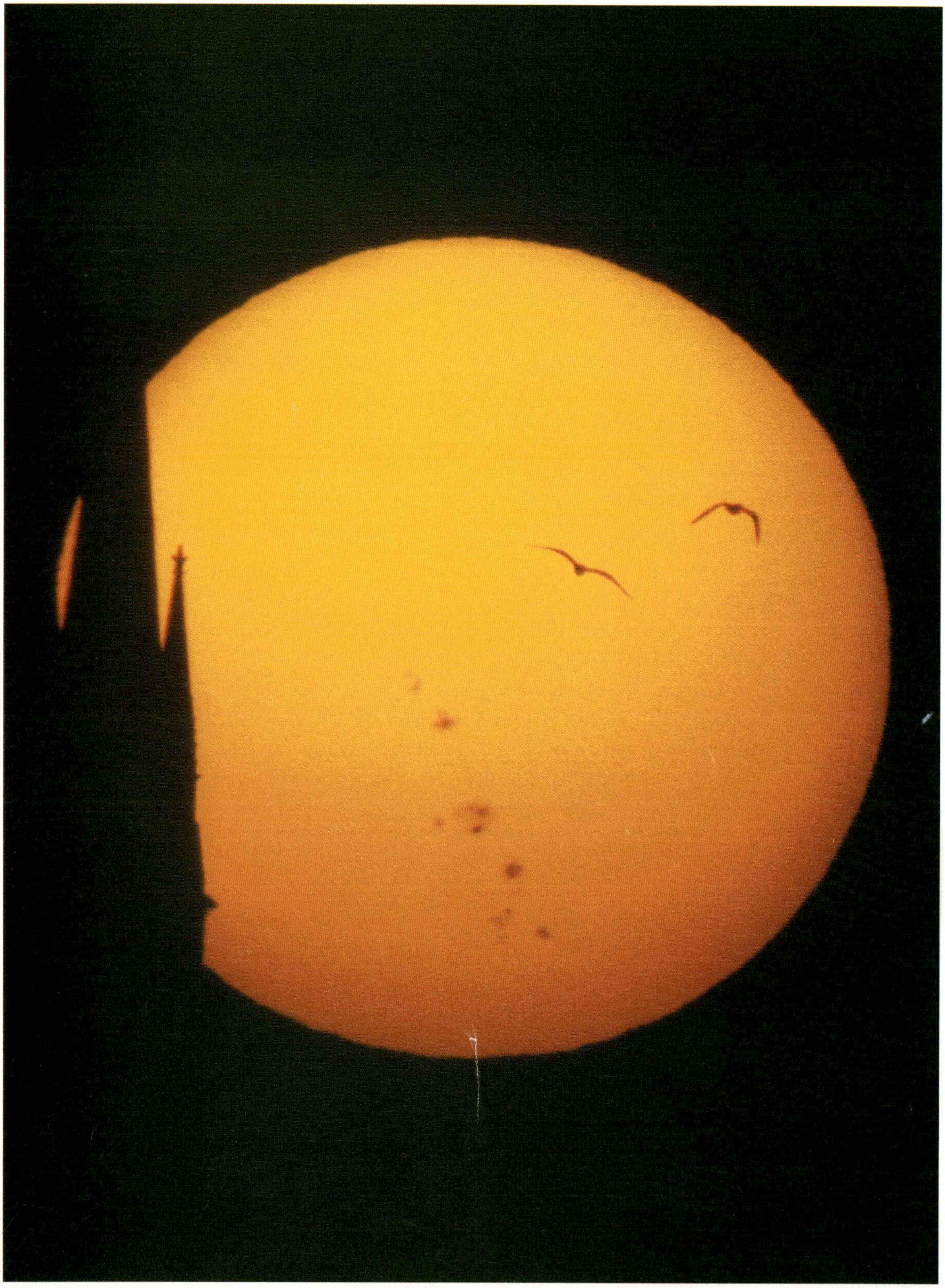

EXPLORING
EARTH IN SPACE

Robert Stephenson and Roger Browne

Illustrated by Jenny Hughes

RAINTREE
STECK-VAUGHN
LIBRARY
The Steck-Vaughn Company
Austin, Texas

Exploring Science

Earth in Space
Electricity
Energy Sources
Forces and Structures
The Human Body
Light
Magnets
Plants
Soil and Rocks
Sound
Uses of Energy
Weather

Library of Congress Cataloging-in-Publication Data

Stephenson, Robert, 1950–
Exploring earth in space / Robert Stephenson and Roger Browne; illustrated by Jenny Hughes.
p. cm. — (Exploring science)
Includes index.
Summary: Presents information about the Earth and its relationship to the sun and other celestial bodies. Includes various activities.
ISBN 0-8114-2603-3
1. Earth—Juvenile literature. 2. Astronomy—Juvenile literature. [1. Earth. 2. Astronomy.] I. Browne, Roger, 1946– II. Hughes, Jenny, ill. III. Title. IV. Series.
QB631.4.S74 1992 91-44198
525—dc20 CIP
AC

Cover illustrations:
Left *The Earth, as viewed from the surface of the moon.*
Right above *An astronomer using the 20-inch Carnegie Double Astrographic Telescope, Lick Observatory, located in Mount Hamilton.*
Right below *A diagram of our solar system.*
Frontispiece *Sunspots can be seen clearly on the lower part of the sun. These spots will "move" as the sun rotates.*

Editor: Elizabeth Spiers
Editor, American Edition: Susan Wilson
Designer: Marilyn Clay
Series designer: Ross George

Typeset by Multifacit Graphics, Keyport, NJ
Printed in Italy by G. Canale & C.S.p.A., Turin
Bound in the United States by Lake Book, Melrose Park, IL

1 2 3 4 5 6 7 8 9 0 Ca 96 95 94 93 92

Contents

The "Blue Planet"	6
Old Ideas	8
The Seasons	10
The Sun's Position	14
Telling Time by the Sun	16
Time Zones	20
Sun and Earth	24
The Moon—Our Satellite	26
The Moon's Surface	28
Eclipses	30
The Solar System	32
The Planets	34
The Sun—Our Star	36
The Galaxy—Our "Star City"	38
Constellations	40
Navigation by the Sun	42
Spaceship Earth	44
Glossary	46
Books to Read	47
Index	48

THE "BLUE PLANET"

No matter what country we come from, what language we speak, whether we live in a town or in the country, near the sea or in the mountains, we all live on the surface of the same slowly spinning planet called the Earth. It is sometimes called the "Blue Planet" because two-thirds of its surface is covered with water.

The Earth is a giant ball nearly 8,000 miles across and 24,860 miles around its widest part. We live on the thin, outer layer, where all life is sustained by air, water, and sunlight. This layer is only from 5 to 25 miles deep. Imagine that the Earth is a model 1 foot across; this life-supporting layer would be as thick as a piece of paper. Some animals live in the depths of the ocean—over a mile below sea level. At about 6 miles above sea level there is not enough air for living things to breathe.

The Earth is sometimes known as the "Blue Planet" because two thirds of its surface is covered with water.

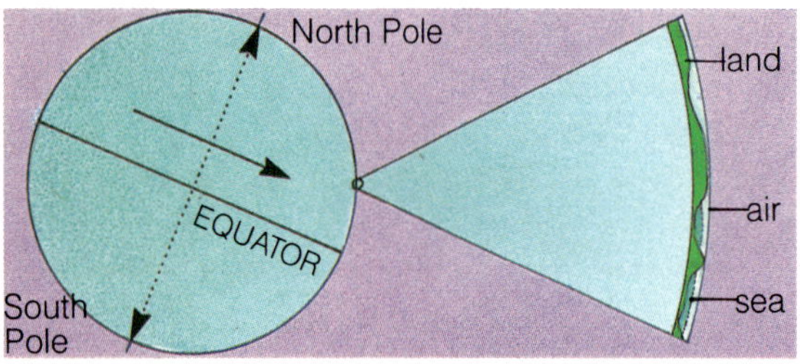

A diagram showing the direction of rotation of the Earth.

Our planet does not just hang motionless in space. It has two important kinds of movement. It spins one complete turn every 24 hours. Seen from above the North Pole, the spin is counterclockwise. We can tell that the Earth spins because the sun "rises" every morning and "sets" every night. During the day, your side of the Earth is facing the sun, and at night, your side is turned away from the sun.

The Earth also revolves around the sun, in a nearly circular orbit (path) that takes just over 365 days to complete. The length of the orbit is 595 million miles. The Earth travels 1.5 million miles every day along its orbit, at a speed of 18 miles per second. We do not notice this speed because the atmosphere travels with the Earth and at the same speed.

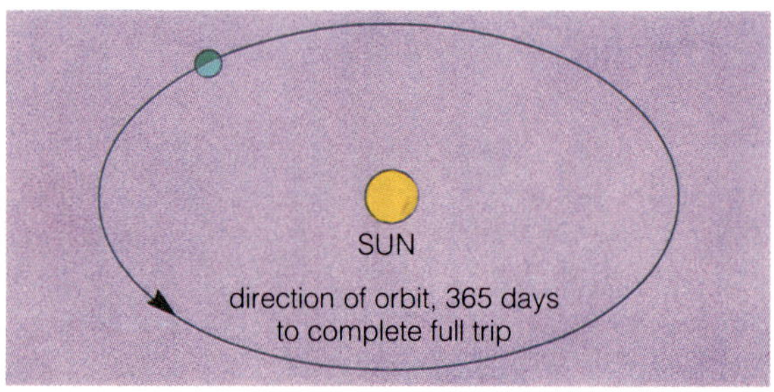

A diagram showing the Earth's orbit around the sun (not to scale).

ACTIVITIES

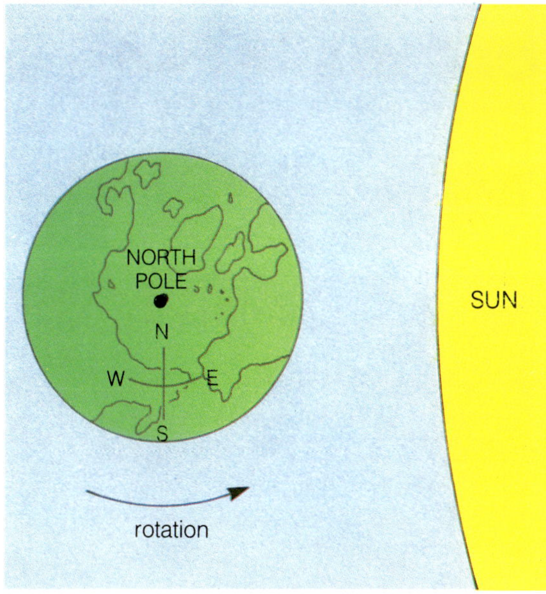

YOU NEED

- **a compass**

1 Study the picture showing the Earth's rotation. Notice which way is east and which is west.
2 From which compass direction would you expect to see the sun appear in the morning? In which compass direction would it disappear at night? Check your answers using the compass.

MODELING THE EARTH'S ROTATION

YOU NEED

- **a dark room**
- **a friend**
- **a flashlight**

1 Stand in a dark room and ask a friend to shine a flashlight on your head. The flashlight represents the sun.
2 Face away from the light.
3 Imagine your head is the Earth. Slowly turn counterclockwise (to your left) until you see the flashlight. Keep turning until it disappears.

4 What happened to the "sun" as you turned?

TEST YOURSELF

1. How thick is the air and water layer on the surface of the Earth?
2. What are the two most important ways in which the Earth moves?
3. How can we tell that the Earth is rotating?

OLD IDEAS

You know that we live on a giant spinning ball called the Earth, which revolves around the sun once a year. This was not always common knowledge. The people of ancient India thought the Earth was a flat disk, held up by four elephants standing on the back of a huge turtle.

Over 2,300 years ago, the Greek philosopher Aristotle discovered that the world is a round ball. He looked at the North Star, which is always seen in the same part of the sky. When he saw it from Greece, it was 40° above the horizon, but when he saw it from Egypt, it was at a height of only 30°. He knew this meant we lived on a spherical (ball-like) planet. The diagram will help you understand.

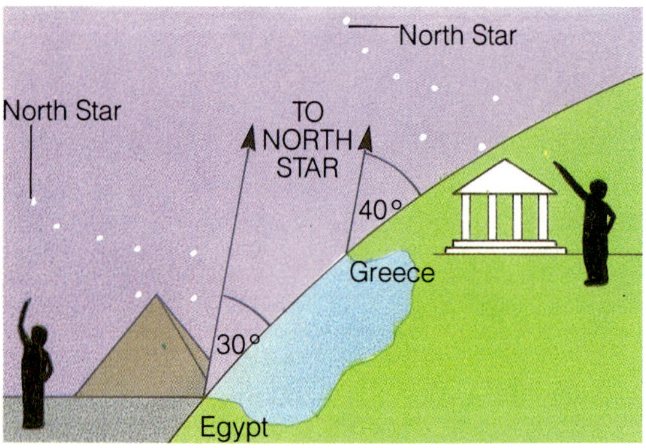

A diagram showing how Aristotle discovered that the Earth is spherical.

Although the ancient astronomers knew the Earth was like a ball in space, they were wrong about its position. They believed that the sun, moon, planets (they had found five of them), and stars moved around the Earth, which was at the center of everything. It is easy to see how they made this mistake. If you do not know that the Earth is spinning, the only way to

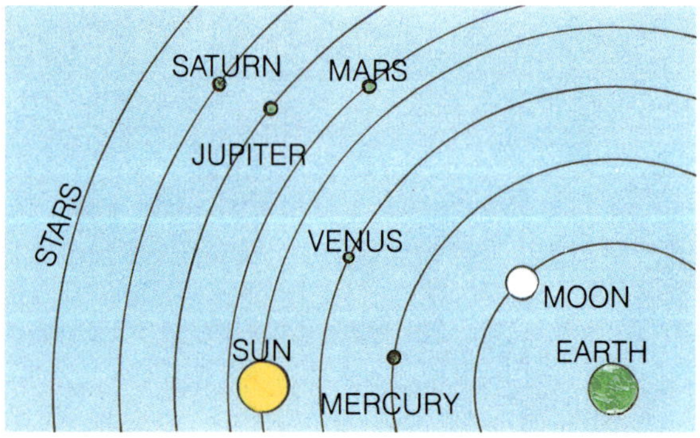

A diagram showing how Aristotle viewed the Earth's position in space.

explain the constant "movement" of the sun, moon, planets, and stars is to imagine that they move around the Earth. It was not until 1543 that the Polish astronomer Copernicus published his idea that the sun was in the center of the universe.

The seventeenth-century Italian scientist Galileo agreed with Copernicus's view that the sun was at the center of the universe.

ACTIVITY

1 There were many other important astronomers at the time of Aristotle and Copernicus. Find out about Ptolemy and why he was so important to our understanding of ancient Greek astronomy. Find out about the discoveries of the Italian scientist Galileo and his troubles with the Church. What did Galileo make that helped him with his observations?

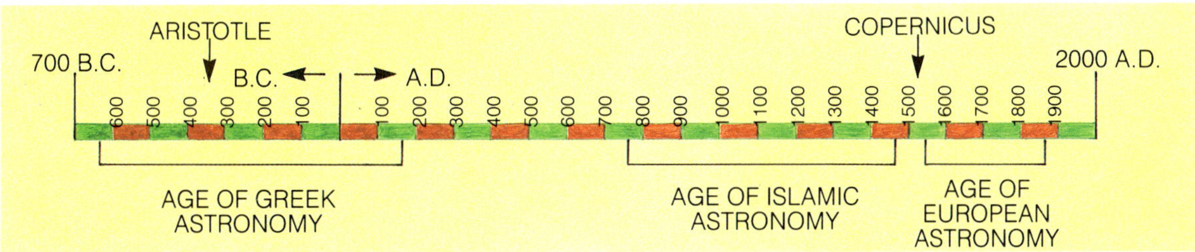

2 Copy the time line above. Add Galileo and Ptolemy. You will add more to it as you read this book.

3 How do the four drawings of the ship steaming toward you help to explain that we live on a spherical planet? Use the information on page 8 to help you with your answer.

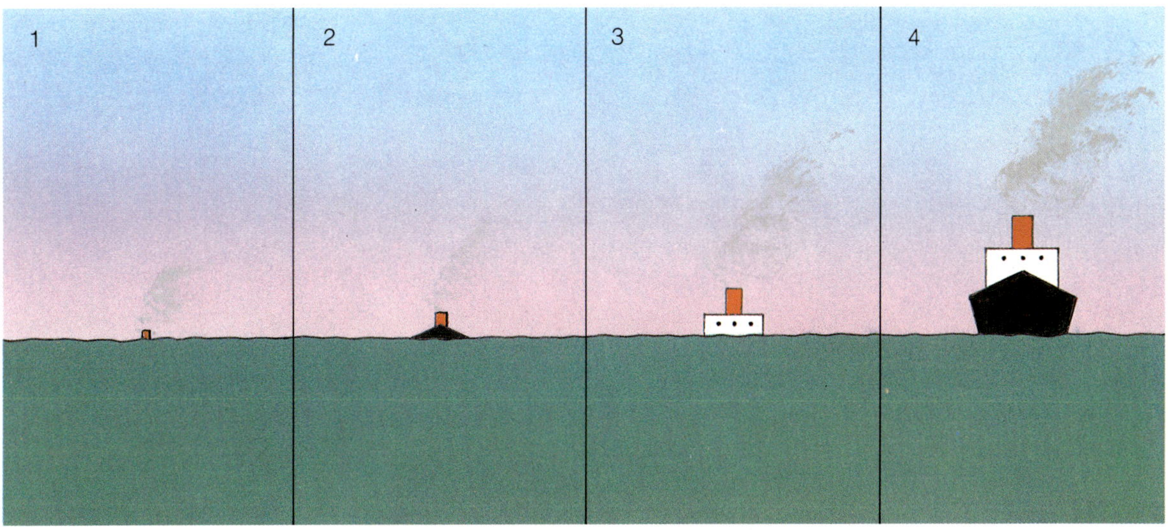

TEST YOURSELF

1. Who discovered that the Earth is spherical?
2. Why did ancient astronomers mistakenly believe that the Earth was circled by the sun, moon, and planets?

THE SEASONS

You know that the rotation of the Earth gives us day and night. Each evening, the rotating Earth carries you out of the light of the sun. At night, you are in the Earth's shadow. The sun rises again as the Earth turns us back into the sun's light.

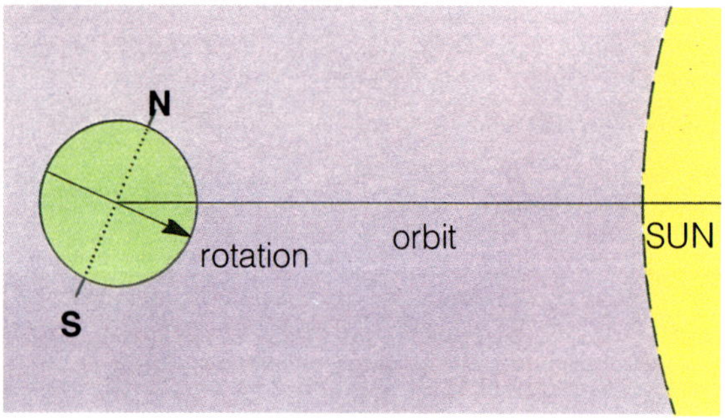

Part of the Earth's orbit, to show the tilt of the Earth.

If the Earth rotated in an upright position, each day would be 12 hours and each night would be 12 hours, all over the Earth. We know this is not true—days are shorter in the winter and longer in the summer. This is because the Earth is not upright in its orbit: it is tilted at an angle of 23.5° from the vertical.

As the Earth travels around the sun, the poles keep pointing at the same places in space. At one point in the Earth's orbit, the North Pole tilts toward the sun and the South Pole tilts away from the sun and is in the Earth's shadow. This is the northern midsummer and southern midwinter. At this time, the North Pole has 24 hours of daylight and there are 24 hours of darkness at the South Pole. Sydney, Australia, has only 8.5 hours of daylight each day and New York City has 16.5 hours. Six months later, it is the South Pole that tilts toward the sun; this is the southern midsummer and northern midwinter. Between summer and winter, the day length gradually

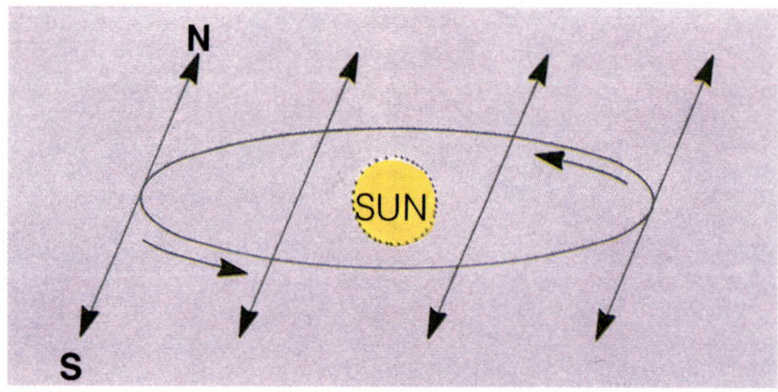

As the Earth travels through its orbit, the poles point to the same place in space.

changes as the Earth moves through its orbit. The tilt of the Earth causes the seasons. The hemisphere pointing away from the sun receives less light and heat. The hemisphere pointing toward the sun receives light more directly and for more hours each day.

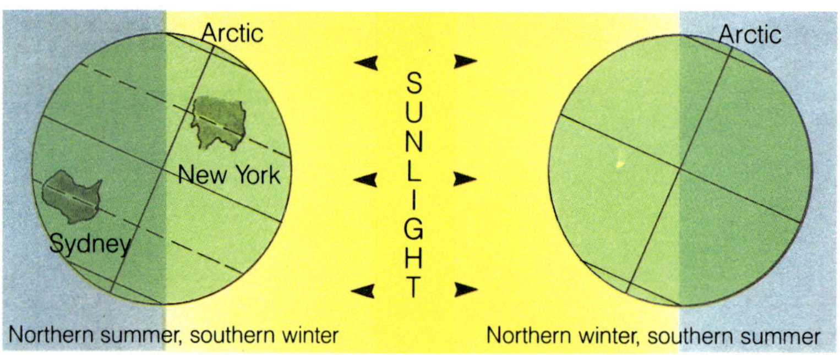

A diagram showing the positions of the Earth in the northern midsummer and northern midwinter, in relation to the sun.

A country landscape in late spring (above) and in winter (below). These seasonal changes are caused by the tilt of the Earth.

ACTIVITIES

A MODEL TO SHOW THE SEASONS

You have seen that the tilt of the Earth gives us seasons and creates the changing number of daylight hours that go with them. A model of the Earth and sun will help you understand how this happens.

YOU NEED

- **a rubber ball with a line drawn for the equator**
- **a thumbtack**
- **modeling clay**
- **a small block of wood**
- **a spotlight or lamp**
- **a large sheet of paper**
- **felt-tip pens**
- **a protractor**
- **cardboard**
- **scissors**
- **an atlas**

1 Cut a wedge of cardboard with an angle of 23.5°.

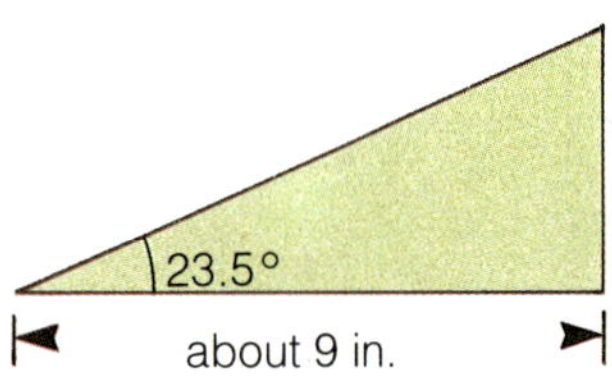

2 Attach the thumbtack to the wood block with clay.

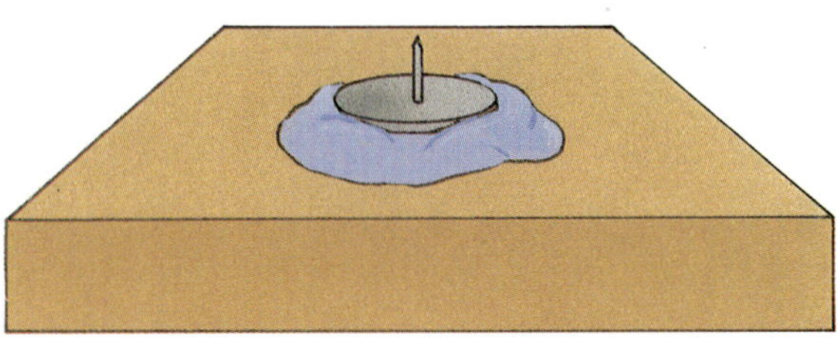

3 Find your latitude in the atlas. This is your distance from the equator, measured as an angle.

4 Use a felt pen to mark your latitude on the ball. Resting one finger on the pole, turn the ball so that the pen draws a latitude line parallel to the ball's equator line.

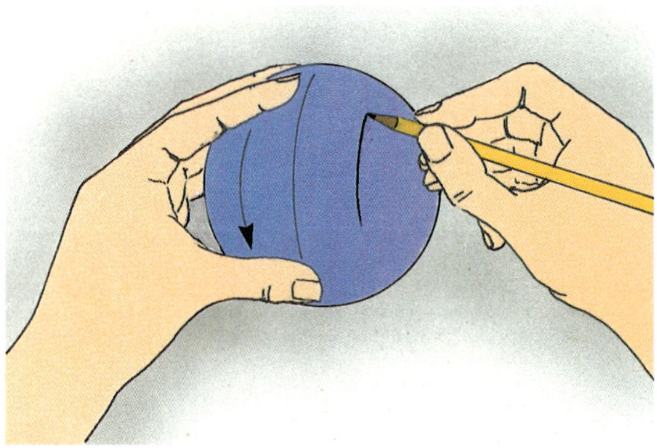

5 Push the ball on the thumbtack so that the equator is parallel to the 23.5° angle of the cardboard triangle.

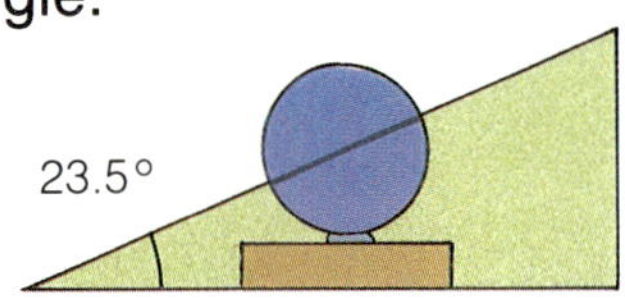

6 Draw a large circle on the paper and mark the center. The circle is the Earth's orbit.

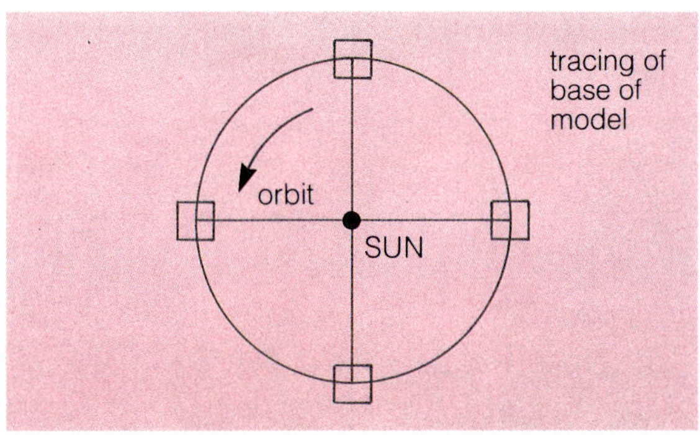

7 Trace around the base of the model Earth, as shown, so that you can move the Earth model around the orbit and keep the tilt of the Earth in the same direction.

8 Place your model in the midsummer (Northern Hemisphere) position. Shine the light on the Earth across the center of the orbit.

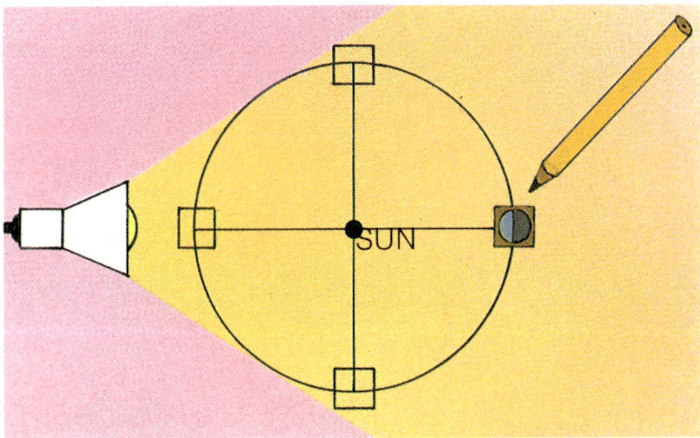

9 The shaded part of your model is night. Notice how much of your latitude line is in daylight.

10 Move the Earth model to each of the four places and draw the night/day boundary line.

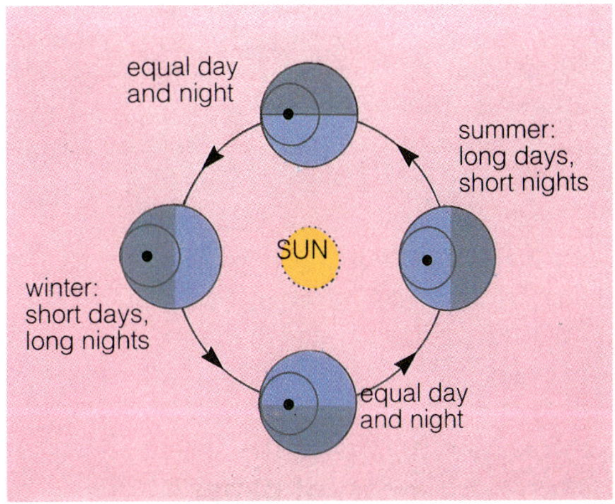

PLOTTING THE HOURS OF DAYLIGHT

1 Use the table of sunrise and sunset to work out the hours of daylight in the middle of each month.

2 Draw a graph of daylight hours for each date in the table.

3 Find two months with 12 hours of day and 12 hours of night.

4 Find the month with the longest daylight hours and the one with the shortest.

Time of sunrise and sunset at 50 °N

DATE	SUNRISE (GMT)	SUNSET (GMT)
mid-January	8:15 a.m.	4:15 p.m.
mid-February	7:15 a.m.	5:15 p.m.
mid-March	6:00 a.m.	6:00 p.m.
mid-April	5:00 a.m.	7:00 p.m.
mid-May	4:15 a.m.	7:45 p.m.
mid-June	3:45 a.m.	8:15 p.m.
mid-July	4:00 a.m.	8:00 p.m.
mid-August	5:00 a.m.	7:15 p.m.
mid-September	5:30 a.m.	6:00 p.m.
mid-October	6:15 a.m.	5:15 p.m.
mid-November	7:15 a.m.	4:15 p.m.
mid-December	8:00 a.m.	4:00 p.m.

TEST YOURSELF

1. Why are there not 12 hours of daylight and 12 hours of night all year?
2. What are the positions of the poles during southern midwinter?
3. Which two months have 12 hours of day and 12 hours of night?

THE SUN'S POSITION

You have seen that the number of daylight hours changes as the Earth orbits the sun (see page 8). This is caused by the tilt of the Earth, which also changes the sun's position during the year. The first of these changes is shown in the diagram opposite.

Diagram A shows midday on June 21, which is the first day of summer in the Northern Hemisphere (the part of the Earth that is north of the equator). The observer is at 52° north of the equator (latitude 52°N) and sees the sun at an angle of 61.5° to the horizon. Diagram B shows midday on December 21 (the first day of winter in the Northern Hemisphere). Here, the observer at 52°N sees the sun at an angle of only 14.5°. So the sun rises higher in the sky in summer than it does in winter. If you look at your shadow on sunny days throughout the year, it will be much longer at 3 p.m. on a winter day than at 3 p.m. on a summer day.

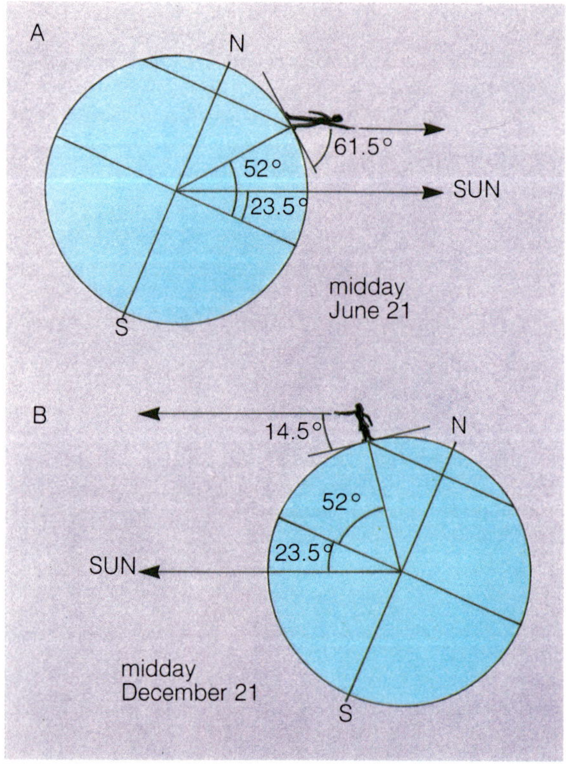

The other important change in the sun's apparent position during the year is where it rises and sets. You know (see page 6) that the sun rises in the east and sets in the west as the Earth turns. This is generally true. However, in summer, the sun rises and sets farther north. In winter, it rises and sets farther south. The diagram below will help you understand why.

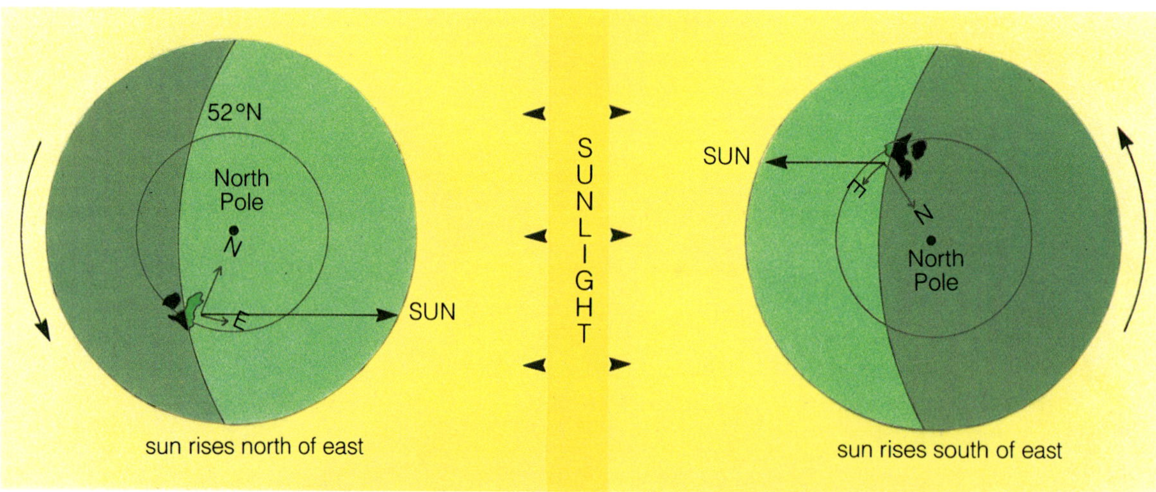

ACTIVITIES

THE POSITION OF SUNSET

YOU NEED

- **sunny evenings**
- **a paper and pencil**
- **a compass**

1 Go outside on a sunny evening and draw the skyline all around you.

2 Use the compass to label your panorama with north, south, east, and west.

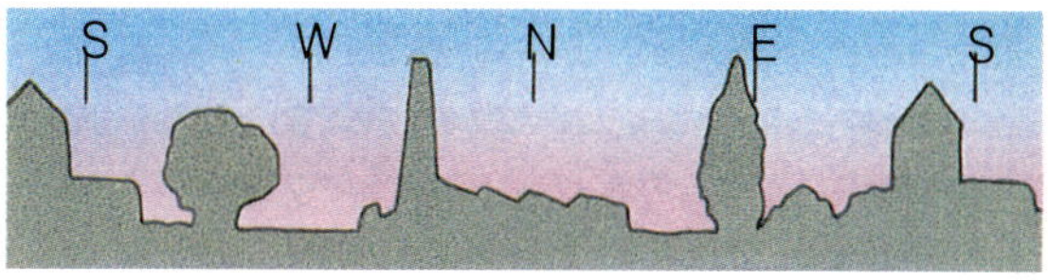

3 Mark and date the position of the sunset on your panorama. Keep adding this information once a month for several months. What do you notice?

CHANGING ANGLE OF THE MIDDAY SUN

YOU NEED

- **an open, level place**
- **a stick about 45 inches long**
- **a ruler**
- **a protractor**
- **paper and pencil**

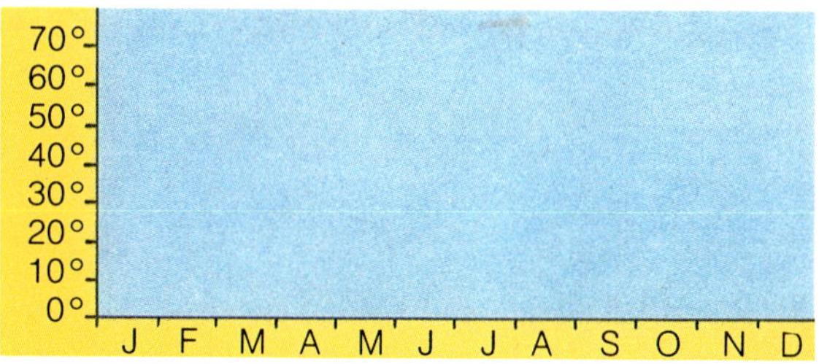

1 Push the stick into the ground so that 39 inches remain above the ground. Use the protractor to check to be sure that the stick is vertical.
2 Once a week at noon, measure the length of the shadow.
3 Use the table to find the angle of the sun above the horizon.

LENGTH OF SHADOW (INCHES)	147	108	84	68	56	47	39	33	28	23
ANGLE OF SUN ABOVE HORIZON	15°	20°	25°	30°	35°	40°	45°	50°	55°	60°

4 Keep a record of the dates and angles to draw a graph. What do you notice?

TEST YOURSELF

1. In what season does the sun rise
- in the southeast?
- in the northeast?

2. How does the angle of the sun in the sky change during the year?

TELLING TIME BY THE SUN

You have learned that the sun "rises" at different places during the year. In the Northern Hemisphere, it appears farther north in the summer and farther south in the winter. Midday is the only time when it appears in the same direction all year round—due south. This knowledge was used in early times to make the first type of sundial, called the noon-dial. This was a rod sticking out from a south-facing wall. The rod is called a gnomon (nō′ men). It was used to show when to ring the church bells at noon. Very early clocks (13th century) were not very reliable, so the noon-dial was also used to correct public clocks.

A sundial that tells the correct time all day must allow for the tilt of the Earth.

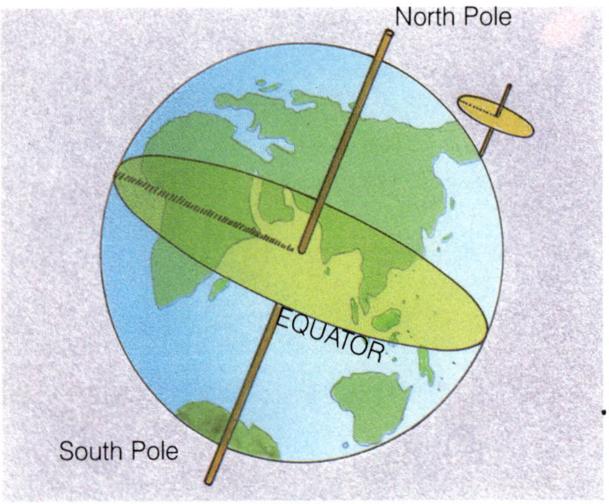

***Above** A diagram showing a model of an equatorial dial, as compared with the tilt of the Earth.*

A noon-dial was used by people to correct their watches and public clocks.

One kind of sundial that does this is the equatorial dial. An equatorial dial is a model of the Earth's equator with the polar axis in place (the polar axis is an imaginary line connecting the North and South poles). Look at the diagram below.

The polar axis of the equatorial dial must be parallel to the Earth's own polar axis. For example, Washington, D.C. is at a latitude of 39°N. A sundial in Washington must be tilted at 39°, as shown below.

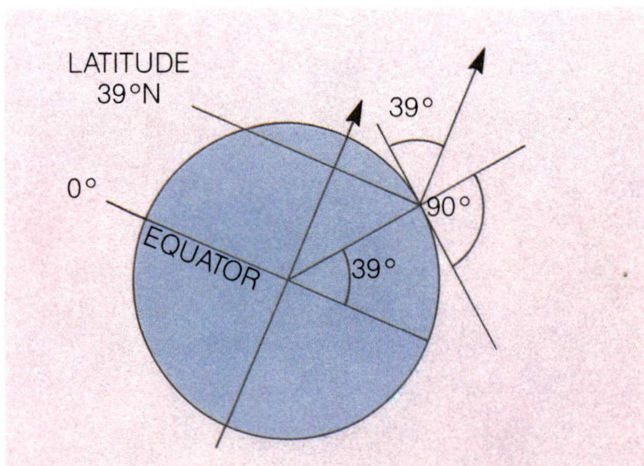

***Right** A diagram showing the tilt of an equatorial dial at the latitude of Washington.*

ACTIVITY

MAKING A SUMMER EQUATORIAL DIAL

YOU NEED

- **cardboard, 6 x 14 inches**
- **a ruler**
- **a pencil**
- **scissors**
- **a gnomon (12 inches of coat hanger wire or stiff wire)**
- **cellophane tape**
- **a protractor**
- **glue**
- **a compass**

1 Crease and fold the cardboard in half.
2 Draw a line down the center on both sides of the cardboard.
3 Measure $4\frac{1}{2}$ inches along the center line from the crease and make a small hole through the board.

An equatorial dial near Greenwich, England. It is tilted to an angle of 51.5°.

4 Repeat step 3, $6\frac{1}{2}$ inches from the fold on the other half of the board.

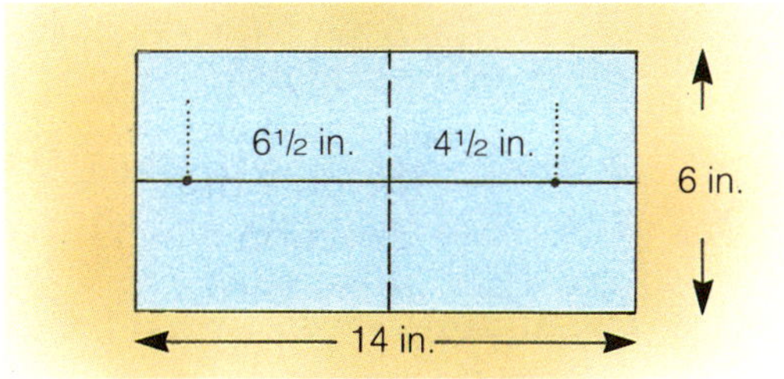

5 Measure $4\frac{1}{8}$ inches from the end of the gnomon. Wind some tape around it to form a stop.

6 Use the protractor to mark the hour lines at 15° from the center line.

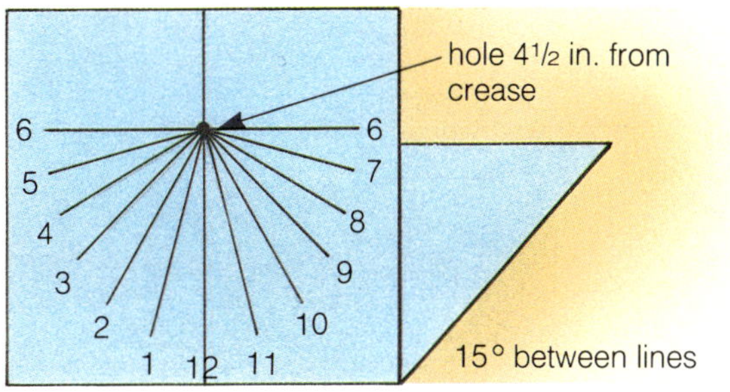

7 Push the long end of the gnomon through the back of the dial face. Fasten the short end in hole A.
8 Glue the back of the face to the stop.

9 To tell time, the gnomon must face north. This sundial is correct at 39°. Change the position of point A and the length of the gnomon to make it correct at your latitude.

GARDEN SUNDIALS

On an equatorial dial, every hour is marked as 15°, because the surface on which it is marked is parallel to the Earth's equator. However, you may be more familiar with the horizontal, or garden, sundial. It is a horizontal, flat plate with the hour lines marked on it. A triangular gnomon is fixed to the plate to cast the shadow. One advantage of the garden sundial over the equatorial dial is that the garden dial can be used to tell the time all year round. However, the dial will work only if it is pointing due north.

Because of the position of the sundial's plate, the hour divisions are not equal. A different dial has to be made for each latitude, and a special device is made to help construct it.

A garden sundial in a park. This type, also known as a horizontal dial, has an advantage over the equatorial dial—it can be used to tell the time all year round.

ACTIVITY

YOU NEED

- **an atlas**
- **cardboard or posterboard**
- **a protractor**
- **a ruler**
- **a pencil or marker**
- **scissors**
- **cellophane tape**

1 Make a right triangle with side *B* 3 in. long; angle y = your latitude; and angle x = 90° – y. Make two copies.

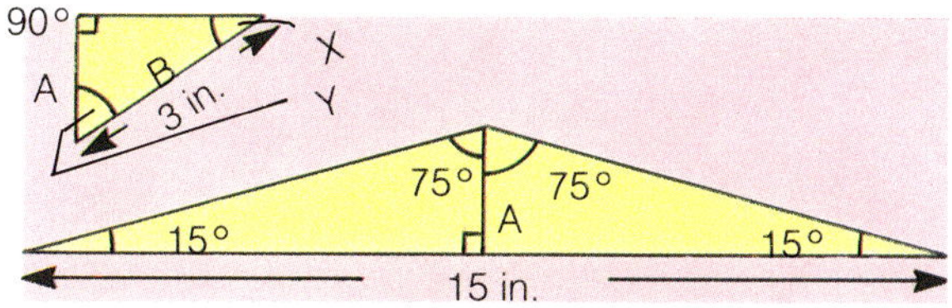

2 Make and cut out a large cardboard triangle as shown. The height equals side *A* of the small triangles.

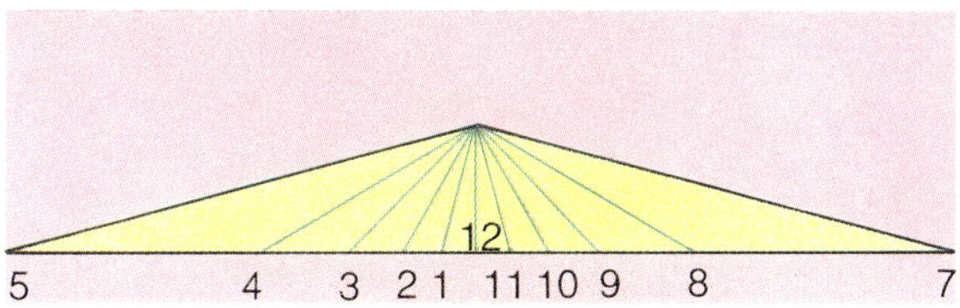

3 Draw lines 15° apart on the large triangle. Mark the hours as shown.

4 Tape a small triangle to the back with side *A* next to height *A*.

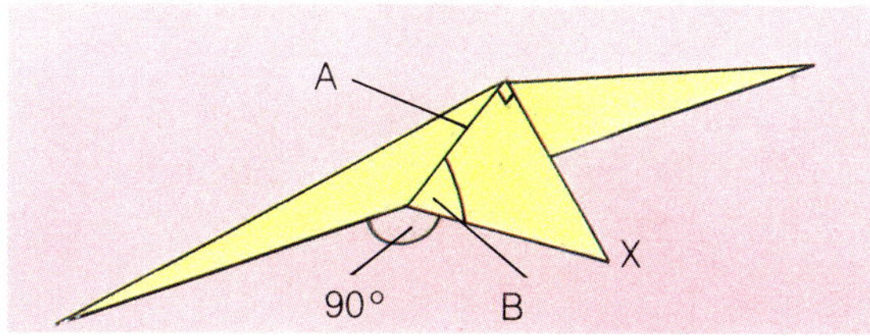

5 Set the triangles, side *B* down, on a piece of cardboard. Mark and label each hour. Mark point x.

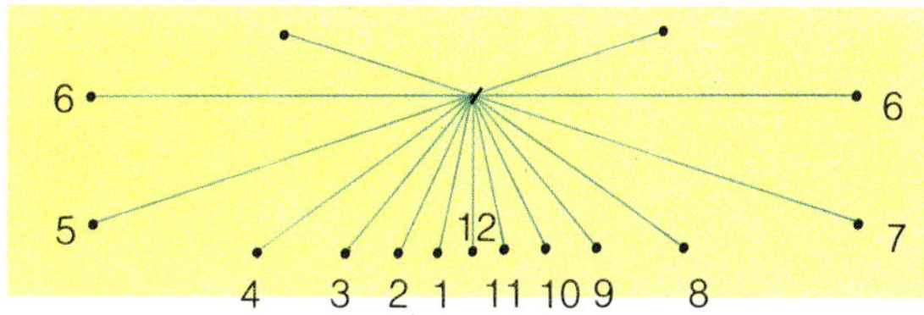

6 Draw lines from point x to each hour. At point x, draw a line parallel to the row of numbers. This is the 6:00 line.

7 Tape a small triangle to the cardboard, x's together and side *B* down. Set the sundial in the sun, facing north.

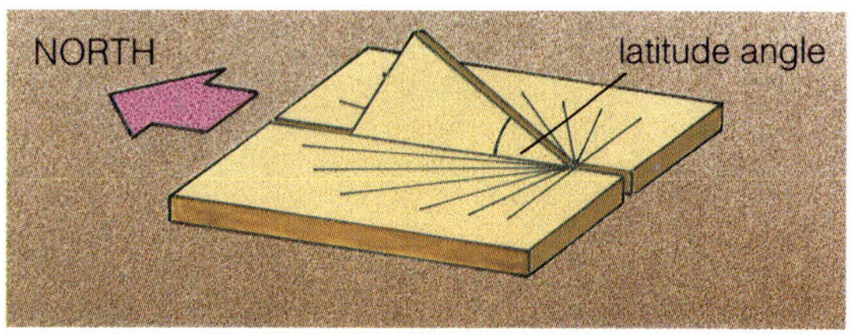

TEST YOURSELF

1. What is a noon-dial? How was it first used?
2. What is an equatorial sundial?
3. What advantage does a garden sundial have over an equatorial dial?
4. How does a garden sundial differ from an equatorial sundial?

TIME ZONES

You know that the Earth makes one complete turn every 24 hours. Because there are 360° in a circle, every place on Earth travels through 360° in 24 hours. This means that any point travels through 15° every hour (360° ÷ 24 = 15°).

Imagine slicing the top off the Earth at the latitude of London (51.5°) to reveal a circle. You could mark the angles around the circle, starting with 0° at Greenwich near London, England. Although there are 360° in a circle, they are shown as 0° to 180° west of Greenwich and 0° to 180° east of Greenwich. Lines drawn through these points, from the North Pole to the South Pole, form lines of longitude.

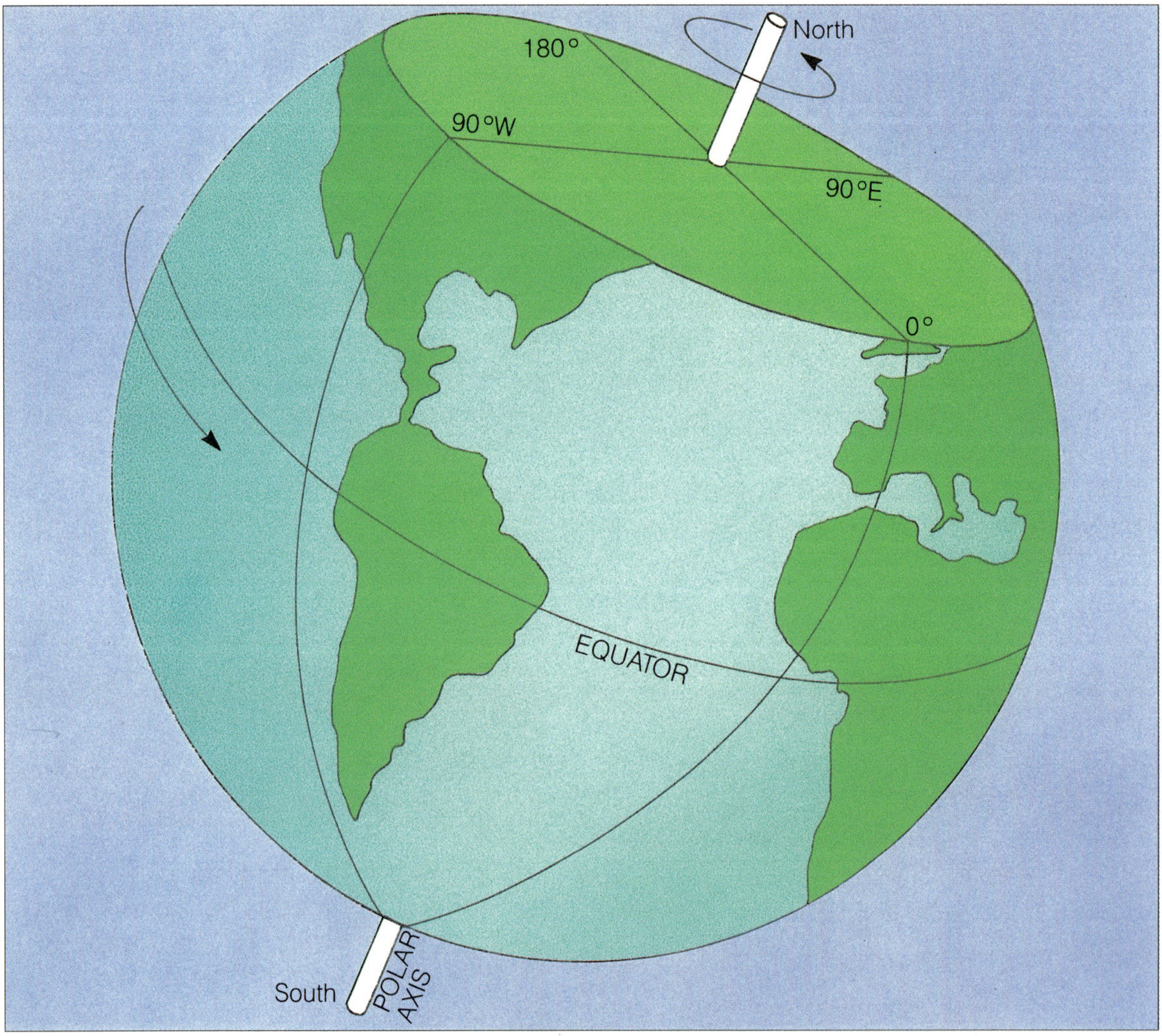

A diagram showing the Earth, sectioned at 51.5° of latitude. The lines of longitude are marked from Greenwich (0°) to 180°E or 180°W.

ACTIVITY

A diagram showing the Earth, seen from above the North Pole on March 21 or September 22 (when the days are exactly 12 hours long).

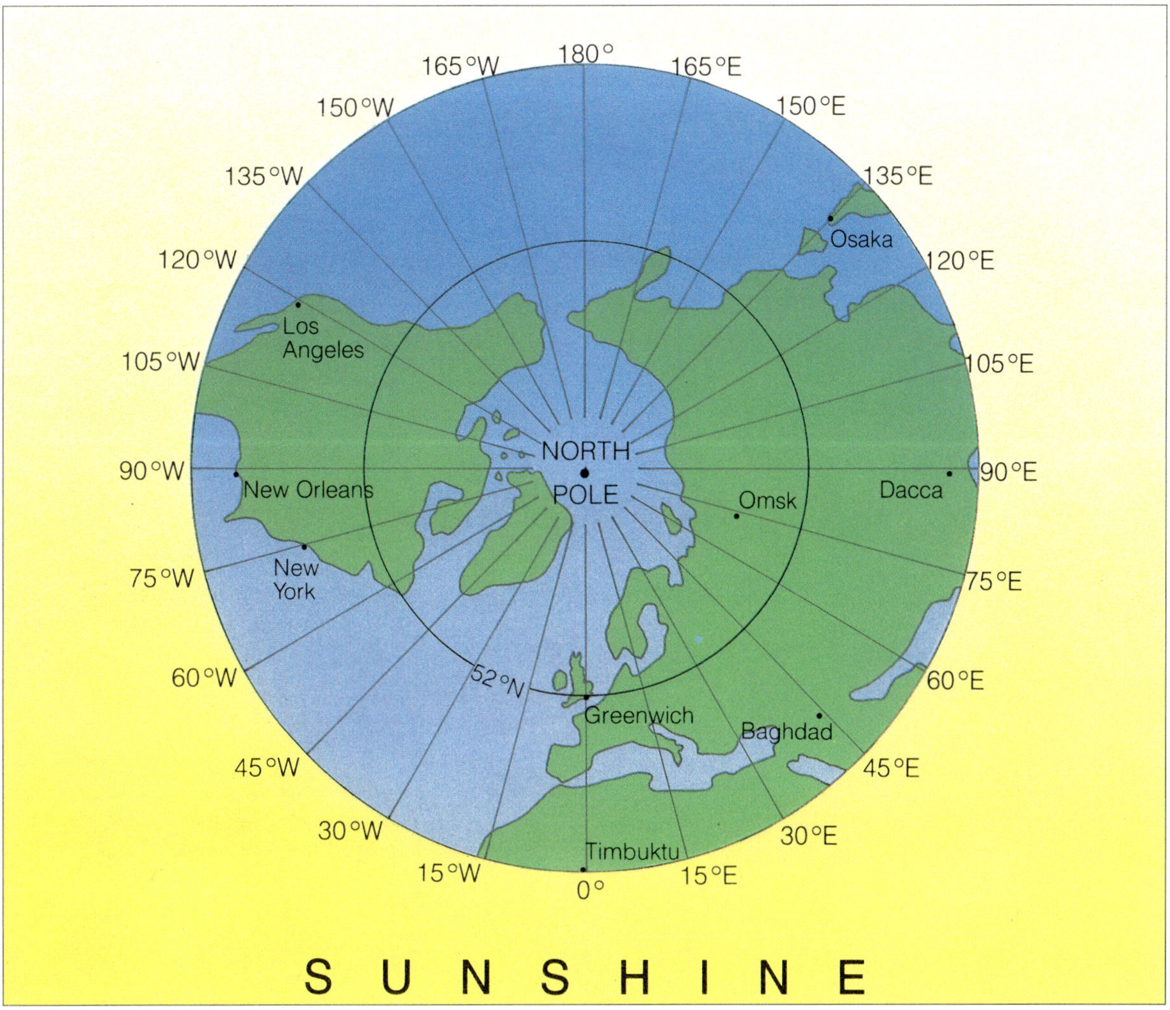

The diagram above shows the Earth from over the North Pole. The sun is at its highest over Greenwich, so it is 12:00 noon in Greenwich.

The date is one of the two times a year when there are 12 hours of daylight and 12 hours of darkness all over the world. Sunrise is at 6:00 a.m. and sunset is at 6:00 p.m.

Each line of longitude shown is at one hour's difference from the next. For example, it is five hours before noon, or 7:00 a.m., in New York City.

What time is it in Timbuktu?
How long is it since the sun set in Osaka?
Is it sunrise, or sunset, in New Orleans?
What is the time difference between Omsk and Baghdad?
When the sun sets in Omsk, how many hours will pass before sunrise in Los Angeles?

You have learned that the Earth rotates through 15° of longitude every hour. If two people live 15° of longitude apart, the person farther west will always be one hour in time (according to the sun) behind the other person. This means that if the person to the east sees that it is 12:00 by his or her sundial, it would be only 11:00 for the person to the west. Every place east or west of another place will have its local time.

Before 1883, different places in the United States had different local times—even in places close to each other. This was not a problem, because most people did not travel far or fast. However, this changed when the railroads came. People started to travel rapidly between towns and to cover greater distances than they had before. They found it very inconvenient to have to put their watches back as they traveled west, or forward as they traveled east. Also, the train timetable at each station was different. Something had to be done to improve this situation. In 1883 standard international time zones were introduced.

The world is now divided into 24 time zones, each one 15° wide. You now have to change your watch by only one hour if you move into the next time zone.

The fastest train in the world—the French TGV (Train à Grande Vitesse). *When the railroads came, standard time zones had to be introduced to avoid confusion.*

When you travel by air, you often have to change your watch by one or more hours as you move from one time zone to another.

ACTIVITY

YOU NEED

- **a large-scale map of your area, showing degrees of longitude**
- **a watch or clock showing Greenwich Mean Time**

1 Using the map, try to find the longitude of your home to the nearest tenth of a degree.
2 Look at the watch. This is the local time at Greenwich (0° longitude).
3 Each degree is equal to 4 minutes of time (60 minutes ÷ 15°).
4 We live west of Greenwich, so take away this time. Those east of Greenwich would add on this time.

TEST YOURSELF

1. What are lines of longitude? How are they labeled?
2. Through how many degrees of longitude does the Earth rotate every hour?
3. How many time zones are there in the world?
4. By how much must you change your watch if your travel through two time zones to the east?

SUN AND EARTH

You have learned that the Earth moves around the sun in a large, almost circular path called an orbit. The distance from the Earth to the sun is also very large. It is, on average, 93 million miles. To show you how far this is, imagine that you could drive from the Earth to the sun in a fast car. If you kept up a speed of 100 miles per hour all the way without stopping, it would take you over 100 years to get there.

The sun is enormous. Its diameter is over 100 times that of the Earth. A million objects the size of the Earth could be packed inside the sun and there would still be space left over.

The sun is not solid like the Earth, but is made up of very hot gases with a surface temperature of 10,000°F. This is about twice as hot as the hottest coal fire. At the center of the sun, the temperature is much higher: around 27 million°F.

The sun is made up of very hot gases. The surface reaches 10,000°F.

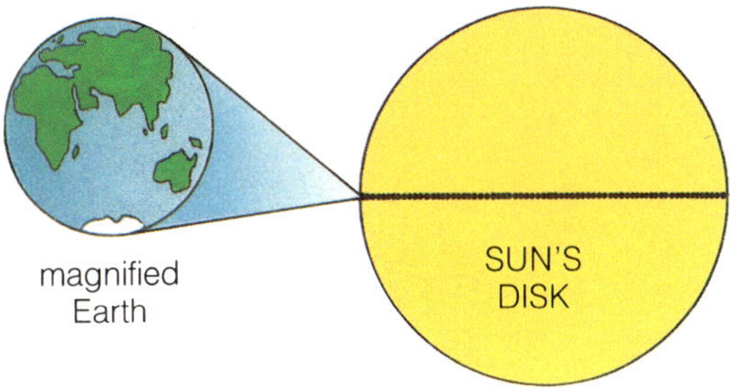

One hundred and nine Earths, placed side by side, would just cross the sun's disk.

The sun does not burn like a coal fire, but produces all its light and heat by changing one kind of gas (hydrogen) into another (helium). This is called a nuclear fusion reaction. This reaction causes the sun to send out large amounts of light and heat, which travel out into space. Without the sun's heat and light, we, and all life on Earth, would not be here.

ACTIVITY

Before moving on, it is worth looking at the sun and earth as a model, just to be sure you have an idea of their sizes and how far apart they are.

YOU NEED

- **a grapefruit (about 4 in. in diameter)**
- **a friend**
- **36 feet of string**
- **a mustard seed or poppy seed**

1 Put the grapefruit in the middle of your playground. The grapefruit is a model of the sun.
2 Ask your friend to hold one end of the string above the grapefruit.
3 Walk away with the other end of the string until it is tight.

4 Hold the seed next to the end of the string. The seed is a model of Earth.
5 Walk slowly around the grapefruit, keeping the string tight. The seed is the Earth orbiting the sun.

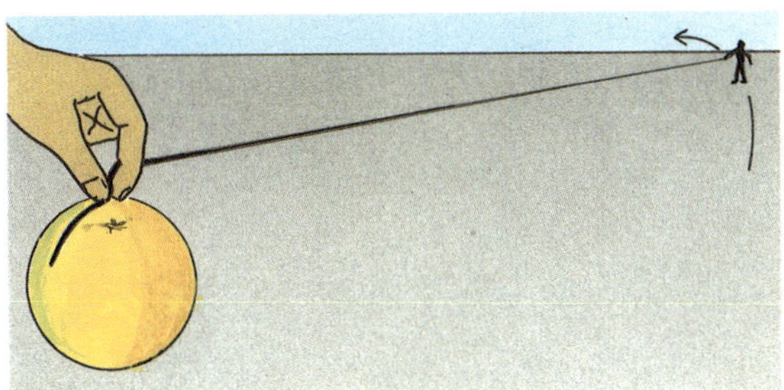

TEST YOURSELF

1. Why is the sun so important to us?
2. If the Earth was a model 1 inch across, how big would a model of the sun have to be?
3. How does the sun produce all its heat and light?

THE MOON—OUR SATELLITE

Just as the Earth travels around the sun, so the moon travels around the Earth. It is the Earth's satellite. It is much smaller in diameter than Earth—only 2,100 miles. The moon is close to us in space. Its average distance from the Earth is 230,000 miles. This is 400 times closer than the sun.

The moon has no light of its own. It shines because the sun lights it up, just as the sun lights up the Earth. Because of this, only half of the moon's surface will have sunlight on it at any time. As the moon moves around the Earth, we see different portions of the sunlit side. This makes it appear to change its shape. These shapes are called phases. The diagram below shows you how the moon appears to change as it moves around the Earth.

It takes the moon 29.5 days to pass through all of its phases. This is known as a lunar month. People who lived thousands of years ago used the moon's phases as a way of telling time.

Although the Earth and moon are close together in space and move around the sun together, they are different in many ways.

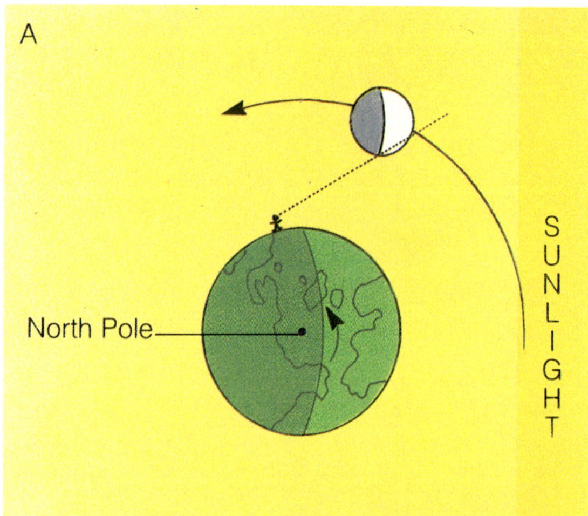

▲ view from space

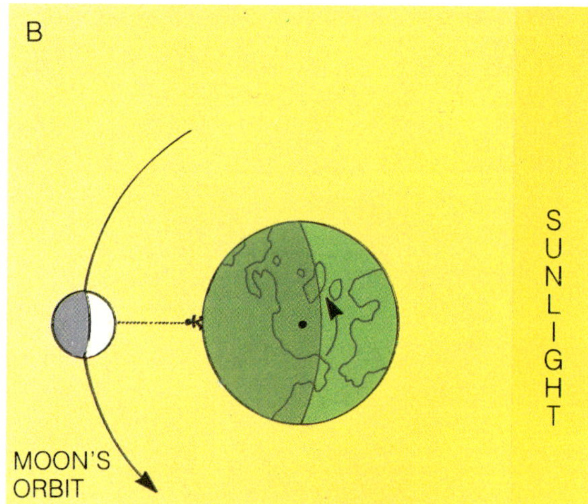

▲ view from space nearly 2 weeks later

Same as A but view from Earth ▼

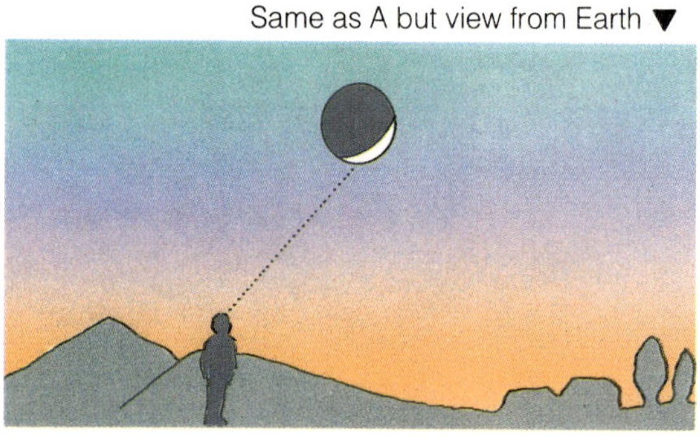

Same as B but view from Earth ▼

Diagrams showing how the moon appears from space and from Earth as it moves around the Earth.

One main difference is their size. This is shown in the picture below.

The moon has no air or water on its surface. It is a silent world covered with craters and mountains, where no life is to be found.

The Earth and moon, drawn to scale.

ACTIVITY

YOU NEED

- **a flashlight**
- **tissue paper (white)**
- **a rubber band**
- **a dark room**
- **2 friends**
- **a light-colored ball (white is best)**

1 Cover the front of the flashlight with tissue paper. This will cut down the glare.

2 Stand in the center of a dark room. You are the Earth.

3 Ask one friend to hold the flashlight about 10 feet away and shine it toward you.

4 Ask another friend to move slowly around you counterclockwise, holding the ball so you can see it clearly.

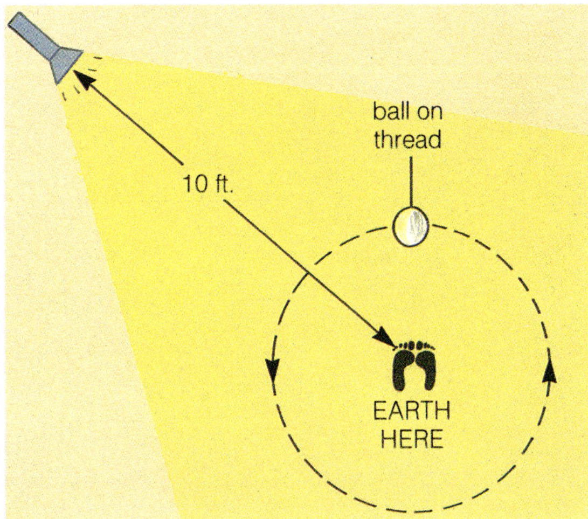

5 Notice how the ball appears to go through phases, just as the moon does.

6 Does your friend holding the flashlight see phases? Try to explain what you have seen.

TEST YOURSELF

1. What is a lunar month? How long is it?
2. As seen from the moon, would you expect the Earth to show phases? Explain your answer.
3. How is the moon different from the Earth?

THE MOON'S SURFACE

On a clear night, especially when the moon is full, you can see bright and dark patches on its surface. The bright patches are mostly highlands and mountains, and they look bright because they are made up of pale rock. The dark patches were once thought to be seas, but are really great, dry plains of dark volcanic rock.

A view of the moon's surface, taken from space by Apollo 11. *Notice the light and dark areas and the enormous number of craters.*

If you use a telescope or powerful binoculars, you will also be able to see the moon's craters. Most of the craters were formed when meteors crashed to the moon's surface. Meteors are large rocks that hurl through space. The moon is covered with craters because there is no atmosphere (air layer). When meteors enter the Earth's atmosphere, most are burned up due to friction between the meteor and the air. This is one of the reasons why the Earth's surface has very few craters.

A crater is a shallow hole with a wall around it. The diagram below shows what a crater looks like from the side. Some craters are very large. The largest crater on the moon is well over 90 miles across.

A side view of a typical lunar crater.

A full moon, as seen from the Earth. This is the best phase of the moon in which to see the surface markings.

The moon is a strange place. It has no air to scatter sunlight, so the sky is always black, even during the day. If you were to stand on the moon's surface, you would see the sun shining out of a black sky. You would also see stars during the day because there would be no lit-up atmosphere to hide them.

The first humans to stand on the moon's surface were Neil Armstrong and Buzz Aldrin. They landed on July 20, 1969, having reached the moon in *Apollo 11.*

ACTIVITIES

LOOKING AT THE MOON

YOU NEED

- **a clear night when the moon is visible (you can find this out from a daily newspaper)**
- **your unaided eyes or binoculars**
- **a sketch pad**
- **a pencil**
- **a flashlight**

1 Look carefully at the moon. Notice the light and dark markings on its surface. The dark markings show up best on a full moon.
2 Draw the moon's surface in as much detail as you can.
3 Display your drawings at school and compare them with those of your friends.
4 Try drawing the moon on different nights when the moon shows different phases. Do you always see the same markings in the same place on the moon's surface? Try to explain your answer.

THE FIRST MOON LANDING

YOU NEED

- **information from page 28**
- **the time line from page 9**

1 Mark the first moon landing on your time line.
2 Find out all you can about the moon landings. Plot their dates on your time line.

One of the astronauts climbing from the lunar module of Apollo 11. *Neil Armstrong and Buzz Aldrin were the first humans on the moon.*

TEST YOURSELF

1. When the sun sets on the Earth, the sky stays light for a little while. This is known as twilight. Why is there no twilight on the moon?
2. What are the moon's "seas"?
3. Why are there few craters on Earth, compared with the moon's surface?

ECLIPSES

From time to time, the moon passes in front of the sun and casts a long shadow that touches the Earth's surface. The diagram shows this happening.

Anyone standing in this "moon shadow" would see the sun disappear. This is called a total solar eclipse and it is very exciting to watch. Total solar eclipses last only a few minutes, but during that time the day becomes like night. Birds stop singing and stars can be seen in the sky. An eclipse of the sun can be seen over only a small part of the Earth's surface.

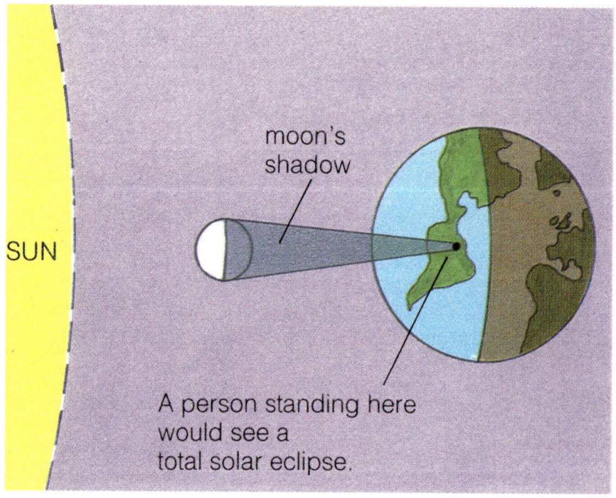

Above *A diagram showing the moon casting a shadow on the Earth during a total eclipse of the sun.*

Left *A total eclipse of the sun.*

This is because the moon casts a small, circular shadow on our planet, which covers only a small area. But this small shadow will travel many thousands of miles across the Earth's surface before the eclipse is over.

For an eclipse of the sun to happen, the sun, moon, and Earth must all be in a straight line. This does not happen every month, because the moon's orbit is slightly tilted compared to the orbit of the Earth around the sun.

When the moon is far above or below the Earth's orbit, the moon's shadow misses the Earth completely, so there is no eclipse. This happens most of the time.

There is another kind of eclipse, called a lunar eclipse (an eclipse of the moon). This occurs when the moon passes through the Earth's shadow, as shown in the bottom picture. Even when the moon is in the Earth's shadow, it is still faintly visible. This is because the air that surrounds the Earth bends some sunlight into the shadow. This gives the moon a reddish-brown color.

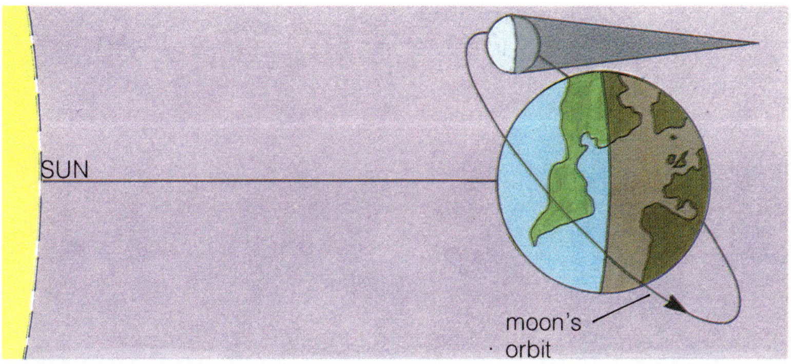

A diagram showing the most common positions of the sun, moon and Earth. The moon is too far above the imaginary straight line between the sun and the Earth, so there is no eclipse.

A diagram showing an eclipse of the moon.

TEST YOURSELF

1. How does a total solar eclipse happen?
2. Why are solar eclipses uncommon?
3. How does a lunar eclipse happen?
4. Why is the moon still faintly visible during a lunar eclipse?

THE SOLAR SYSTEM

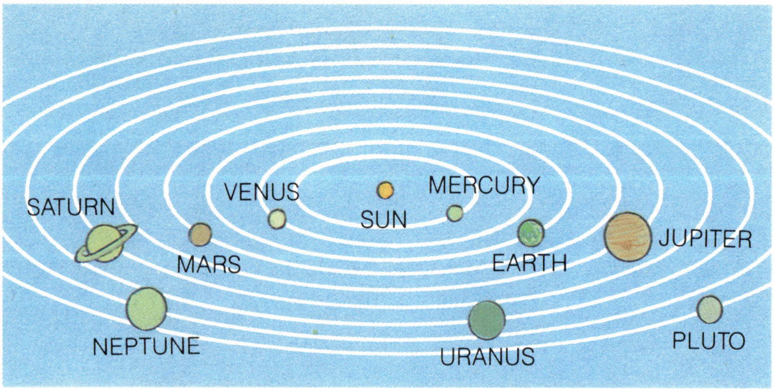

A diagram of the solar system. All the planets move around the sun in the same direction.

You have learned about the sun, moon, and Earth. You know that the moon travels around the Earth once every month, and that the moon and Earth travel together around the sun once every year.

But there are other planets that also travel around the sun. These planets, along with other objects called comets and asteroids, make up what we call the solar system. This is the sun and everything that travels around it.

Five of the planets are bright enough to be seen with the unaided eye. This is why they were discovered by people in ancient times. They are Mercury, Venus, Mars, Jupiter, and Saturn. You can spot them quite easily if you know where to look. However, they are not visible all the time because they occasionally move into our daytime sky. Do not expect to find them marked on star maps because the planets do not remain in any fixed part of the sky.

The William Herschel telescope in La Palma, Mexico. Large telescopes are needed to see Pluto, because it is so far from Earth. Large telescopes can also give astronomers much information about other planets.

However, astronomy magazines publish planetary positions monthly, so you can use one of these to spot your first planet. Venus is very easy to find, because it appears as a brilliant object in the evening sky after sunset for a few weeks, roughly every eighteen months.

The other three members of the solar system are too dim to be seen with the unaided eye, because they are so far away from Earth. They are Uranus, Neptune, and Pluto. Uranus and Neptune can be seen with binoculars, but Pluto is so faint that a large telescope is needed to see it.

ACTIVITY

A MODEL OF THE SOLAR SYSTEM

YOU NEED

- **9 friends**
- **a playground or field**
- **a tape measure**

1 One person will be the sun and the rest will be the planets.
2 The person who represents the sun stands in the middle of the playground or field.
3 Each person chooses a planet. Using the list below, work out how far he or she must stand from the sun. (Let 1 inch equal 1 million miles.)

Distance from sun (millions of miles)

Planet	Distance	Planet	Distance
Mercury	36	Jupiter	480
Venus	67	Saturn	900
Earth	93	Uranus	1800
Mars	141	Neptune	2800
		Pluto	3600

4 Each person uses the tape to measure his or her distance from the sun. Check your answers with your friends to make sure that all the distances are correct.
5 Take your positions.

6 What do you notice about the distances of the four planets closest to the sun?
7 Are the orbits in the solar system evenly spread out?
8 Which planet would you expect to take the longest time to travel around the sun? Which would you expect to take the shortest time?

TEST YOURSELF

1. Why can you not always see the bright planets?
2. Why are the planets not marked on star maps?
3. Which planets cannot be seen with the unaided eye?

THE PLANETS

You have seen how the planets move around the sun. The ancient Greeks noticed the movements of five of them against the background of stars and called them planets, which means wanderers. We now know of a total of eight planets besides Earth in the solar system.

Mercury, the closest planet to the sun, is a little larger than our moon. Like the moon, Mercury has no atmosphere and is covered with craters. It has no satellites (moons).

Venus, next to Mercury, is almost the same size as Earth, and has been called Earth's twin, although the two planets are very different. Venus is completely covered by clouds, and its atmosphere contains a vast amount of carbon dioxide gas. This has caused a very strong greenhouse effect, making Venus's surface too hot for life to exist. Venus is thought by many scientists to provide a lesson for us on Earth: an example of what might happen if we continue to release large amounts of carbon dioxide into our atmosphere.

Mars, the famous "Red Planet," lies beyond Earth's orbit. Mars is just over half the size of Earth and has two satellites. It was once thought that life might exist on Mars and in 1976 two space probes went there to look, but no life was found.

Jupiter is the giant of the solar system. It is covered by clouds and is striped. But its best-known feature is the Great Red Spot, thought to be a huge, permanent storm. Jupiter has sixteen satellites, some of which can be seen through powerful binoculars.

Saturn, the famous ringed planet, is a beautiful sight through a telescope. It is slightly smaller than Jupiter, but twice as far from Earth. It has seventeen satellites.

Uranus is smaller than both Jupiter and Saturn and was the first planet to be discovered using a telescope. It was discovered in 1781 by the amateur astronomer William Herschel, using a homemade telescope in his backyard. Uranus has fifteen satellites. In 1977, it was discovered to have rings.

Neptune was discovered in 1846, but not by accident. Astronomers began looking for it when they discovered something was "pulling" Uranus out of position. The planet was finally found by two German astronomers—Galle and D'Arrest.

Pluto, thought to be the outermost planet, is a mystery. It is as small as our moon and may not deserve the title of planet, as it seems to be nothing like any of the other planets. It was found in 1930 by

A diagram showing the planets of the solar system to scale.

the astronomer Clyde Tombaugh, who discovered it on three photographs that he took through a large telescope. In 1978, it was found to have a satellite. With the discovery of Pluto, the size of the solar system was increased by over four times that which was known to the ancient astronomers.

ACTIVITIES

YOU NEED

- **the time line from chapter 2**

Plot the discoveries of Uranus, Neptune, and Pluto on your time line.

FINDING OUT ABOUT THE SOLAR SYSTEM

Astronomers have found out a lot about the other planets in the solar system. Many of the new discoveries have come from space probes, such as *Voyager I* and *II*. You can find out about these for yourself from the many new astronomy books and magazines that are available.

You can also find out about other objects in the solar system, such as asteroids and comets.

YOU NEED

- **up-to-date books on the planets and solar system**

1 Work as a group. Each person should choose a planet (or more than one, if you are in a small group).
2 Find out as much as you can about your chosen planet.
3 Find out about the other objects in our solar system, such as asteroids and comets.
4 Make a group display at school, showing your findings.

This photograph of Saturn was taken by one of the Voyager space probes. These spacecraft have given astronomers a great deal of information about our solar system.

TEST YOURSELF

1. What does the word planet mean?
2. Which planets were known before telescopes were invented?
3. Name the largest planet in the solar system.

THE SUN—OUR STAR

You have learned that the sun is a large ball of hot gas that gives off heat and light. It is so bright that it is very dangerous to look at it directly. Astronomers often examine the sun by projecting its image onto a screen. Viewed in this way, the sun has dark spots on its surface—they are called sunspots. They are caused by strong magnetic fields on the sun's surface.

When you stand outside on a clear, dark night and look out at the stars, you are looking at other "suns" far away in space. Our sun appears so large and bright only because it is so much closer to us. Some stars are really much bigger and brighter than the sun, and appear dim only because they are so much farther away. Our star is, of course, very important to us, but it is quite small and ordinary when compared with some other stars.

A diagram showing our sun and some other stars to scale, with their surface temperatures.

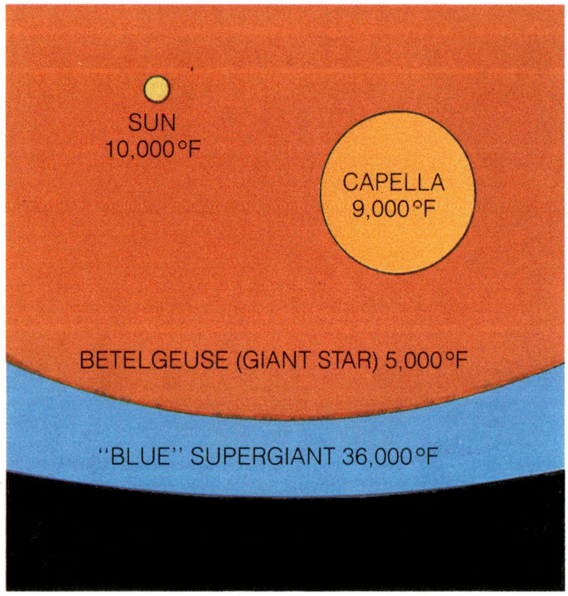

The distances between the stars are very great. If you made a scale model of the solar system, with the sun the size of a ping-pong ball, then Pluto, the farthest planet, would be a speck of dust 500 feet away. The nearest star (Proxima Centauri) would be another ball 600 miles away! It can be seen from this that the solar system is very small compared with the distances between the stars. Some astronomers think that there might be planets orbiting other stars. If this is so, then life might exist on some of these planets. Imagine an alien on some distant planet looking up at the night sky. For that alien, our sun would be just another small star shining in the night.

A sunspot, photographed by the space shuttle in the summer of 1985. Sunspots are much cooler than the normal surface temperature of the sun.

ACTIVITY

LOOKING FOR SUNSPOTS

YOU NEED

- **binoculars with a tripod or stand**
- **two pieces of white cardboard**
- **a pencil**
- **an adult to help you**
- **a sharp knife or utility knife**
- **cellophane tape**
- **a ruler**

1 Set the binoculars on a piece of cardboard and draw outlines of the lenses.

2 Ask an adult to cut out the outlines. Do not attempt to do this yourself.

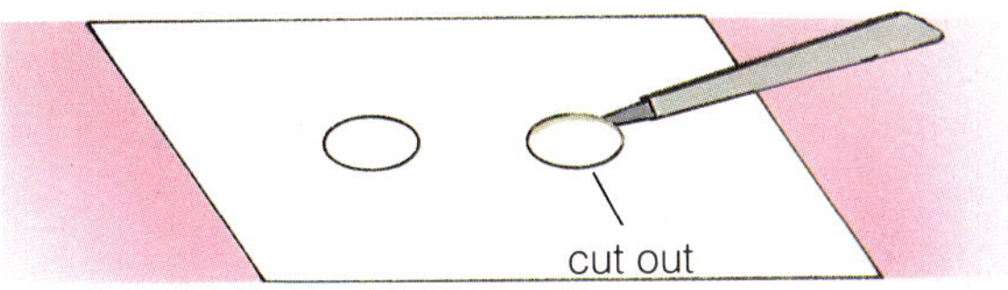

3 Lining up the holes with the lenses, tape the cardboard to the binoculars.

WARNING: Never look directly at the sun with or without binoculars. It is very dangerous and could make you permanently blind.

4 Attach the binoculars to a tripod or stand.

5 Ask an adult to point the binoculars at the sun.

6 Place the other piece of cardboard about 18 inches behind the eyepiece. Project the sun's image onto it.

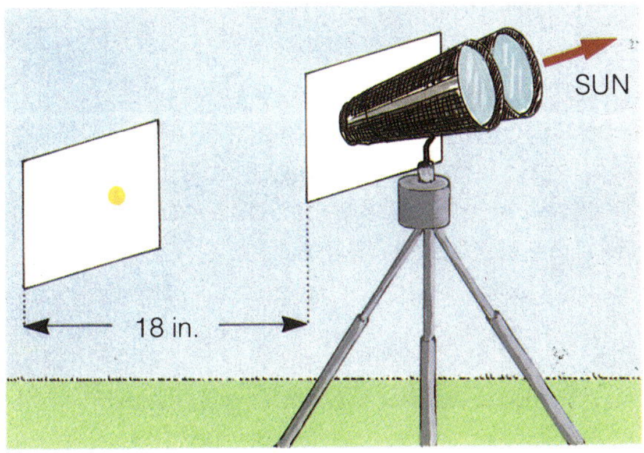

7 Adjust the focus until the sun's image appears sharp on the card.

8 Look carefully at the image. Can you see any dark markings on the sun's image? If you can, mark their positions on the card.

9 Follow the sunspots every day. You will see that they gradually move across the surface. This is because the sun, like the Earth, is rotating on its axis.

TEST YOURSELF

1. What is the difference between a star and a planet?
2. Why do sunspots move?
3. Why do astronomers often project the image of the sun onto a screen?

THE GALAXY—OUR "STAR CITY"

If you stand outside on a dark summer's evening, well away from artificial light, and look toward the south, you will see a faint, cloudlike band of light stretching up from the horizon. This faint band of light is called the Milky Way.

Part of the Milky Way—our galaxy—as seen from Earth. The white line is the trail caused by a satellite.

If you look carefully, you will see that the Milky Way stretches across the sky. Our sun is one of the millions of stars that make up this galaxy. The faint band of light is caused by the glow of distant stars, all lying along a narrow region of sky.

Our galaxy is like a great "star city" floating in space. The solar system lies a long way from the center, out in the "suburbs" of this city. It is very hard to imagine the size of our galaxy, the Milky Way. Astronomers have measured its size, but they do not use miles; instead, they measure these enormous distances in light-years. A light-year is the distance a ray of light travels in one year. Light travels very fast; if you switch on a flashlight, the beam will move away from you at 180,000 miles every second. At this speed, it would take just over a second to reach the moon, and only eight-and-a-half minutes to reach the sun. But, even at this speed, it would take the beam of light 100,000 years to travel across our galaxy.

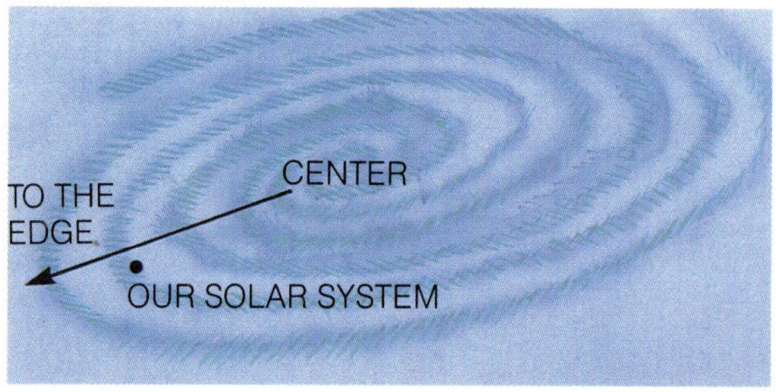

A diagram showing how our galaxy would appear from the outside.

Astronomers using large telescopes have discovered millions of other galaxies far beyond our own. The closest one is called the Andromeda Galaxy. It is the farthest object we can see with the unaided eye. It is worth trying to catch a glimpse of this faint object, for the light you see has had to travel for two million years to reach us.

The Andromeda Galaxy is the closest galaxy to our own. It is the farthest object we can see with the unaided eye.

ACTIVITIES

FINDING THE MILKY WAY

YOU NEED

- **a clear view to the south**
- **a clear, dark night (August or February are the best months)**
- **a place far from street lights**
- **a moonless night**
- **binoculars**
- **a star map**

1 Look toward the south around 9:00–9:30 p.m.
2 Allow at least 10 minutes for your eyes to adjust to the dark.
3 Try to pick out the faint band of light stretching up from the horizon. It will look like a long, ragged cloud.
4 Follow it with your eyes, up over your head and down toward the northern horizon.
5 Try to explore it with binoculars. You will see many faint stars and glimpse clouds of hydrogen gas.
6 Use a star map to find your way around the Milky Way. Check what you have seen from the star map and learn more about the stars.

FINDING THE ANDROMEDA GALAXY

YOU NEED

- **a star map**
- **an evening in autumn**
- **a dark, clear, moonless night**
- **a place far from street lights**

1 Find the Square of Pegasus on your star map. Find it in the sky.
2 Look at the drawing below. This shows you where to look for the Andromeda Galaxy.

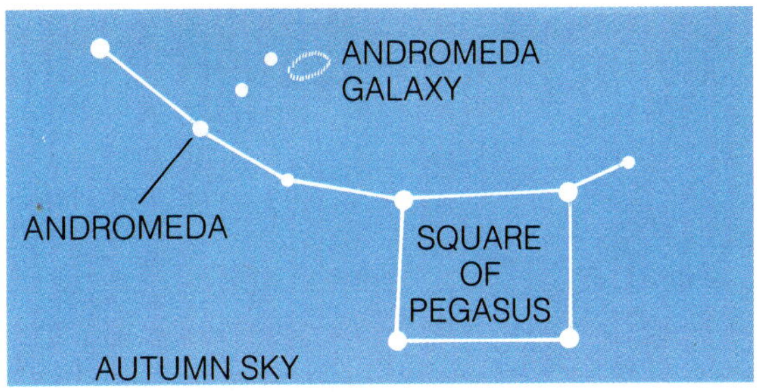

3 Allow 10 minutes for your eyes to adjust to the dark. The Andromeda Galaxy will appear as a faint, fuzzy blob.
4 Try looking just to the side of it—that will make it easier to see. Think about how long the light has had to travel before it reaches your eyes.

TEST YOURSELF

1. Is the solar system closer to the center, or to the edge, of our galaxy?
2. What is the Milky Way?
3. If you could travel at the speed of light, how long would it take you
- to travel to the sun?
- to travel across our galaxy?

CONSTELLATIONS

A diagram showing some of the main constellations that appear at different times of the year.

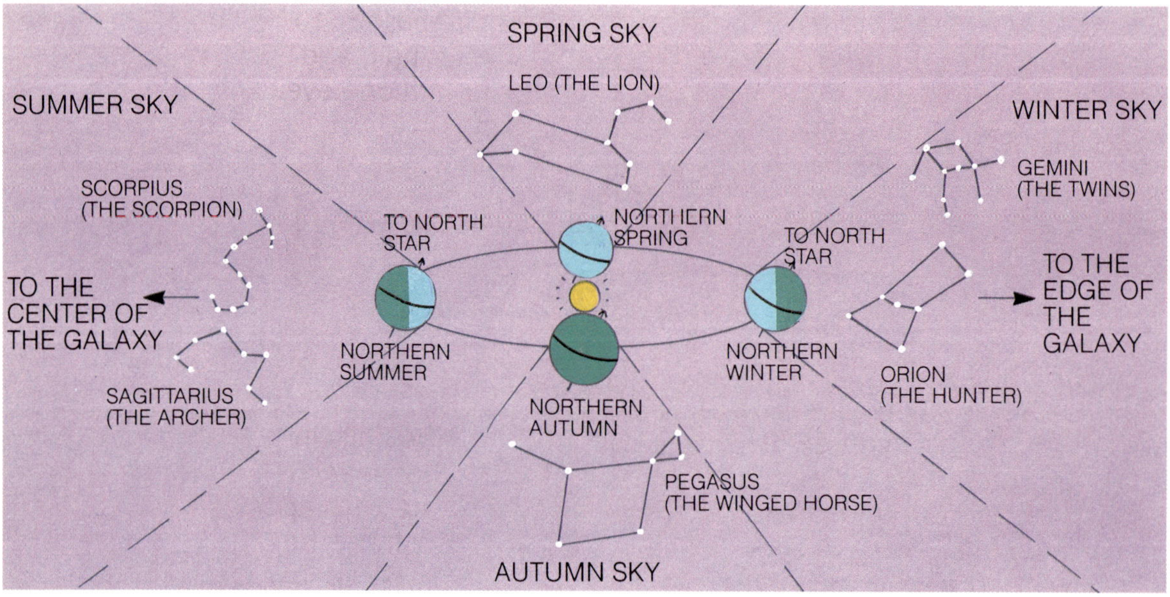

A constellation is the name astronomers give to a group of stars that seems to form a pattern. Many of the constellations we see today were named by people thousands of years ago. The ancient astronomers thought that some star patterns resembled heroes and creatures from their legends. Only a few constellations really look like the things they are supposed to represent. It seems that the ancient astronomers had very good imaginations!

The constellations help us find our way in the sky. As the Earth travels around the sun, we see different constellations at different times of the year.

In summer, the night side of the Earth is turned toward the center of our galaxy. When we look in that direction, we see constellations such as Scorpius and Sagittarius. In winter, our night sky faces the other way. Therefore, we see the constellations lying in that direction, such as Orion and Gemini.

Most of the stars belonging to a constellation lie at different distances from us and from each other, so constellations do not really exist in space. They are just the patterns formed by stars lying in the same direction, as seen from the Earth.

All the stars we can see with the unaided eye belong to our galaxy. The closest star to the solar system is Proxima Centauri, which is 4.3 light-years away. This means its light takes 4.3 years to reach us.

You have learned that our galaxy is only one among millions of other galaxies. These galaxies make up what we call the visible universe. The universe is everything that exists. Astronomers have discovered that the galaxies are moving away from each other, so the universe is gradually spreading out, or expanding.

ACTIVITIES

MAKE A CONSTELLATION

YOU NEED

- **5 or more polystyrene balls**
- **5 or more differing lengths of stiff wire**
- **a table or other flat surface**
- **a dark-colored piece of cardboard**
- **modeling clay**

1 With the help of an adult, stick the end of each piece of wire into each polystyrene ball.
2 Stick the other end of each wire into some clay to form a base.
3 Set up the balls on the table or other flat surface. Arrange them so that they are not all in a line.

WARNING: Ask an adult to help you stick the wires into the polystyrene.

4 Place the dark card behind the balls so that they are easier to see.
5 Move about 6–10 feet from the nearest ball, and view the pattern of balls with one eye. Draw what you see.

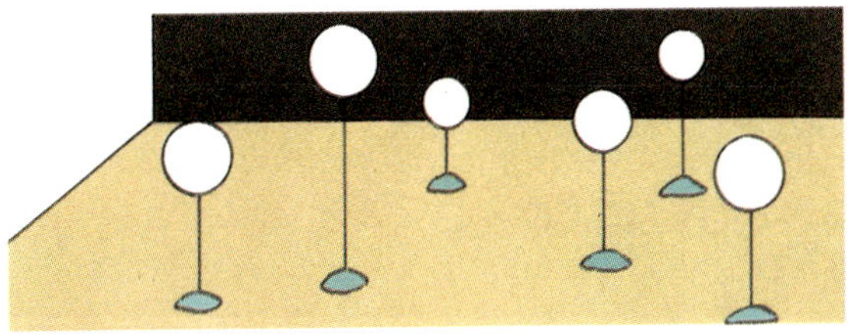

6 Move the balls into different positions on the table and repeat the activity.
7 Do all the balls appear to be at the same distance from you? What does this tell you about constellations?

LOOKING AT SOME CONSTELLATIONS

YOU NEED

- **a map of the night sky**
- **a dark, clear, moonless night**
- **a dim flashlight**
- **a paper and pencil**

1 Look at your star map and find the North Star.
2 Find the constellations Cassiopeia and Ursa Major. They are close to the North Star. Both are easy to spot and are above the horizon all year round for people living above latitude 50°N. Cassiopeia looks like a W or an M and Ursa Major looks like a saucepan.
3 Draw these star patterns. Notice how they move around the sky during the year.

TEST YOURSELF

1. What is a constellation?
2. Why do constellations not really exist in space?
3. Why do we see different constellations at different times of year?

NAVIGATION BY THE SUN

Sailors often travel for days without spotting any landmarks, so it is important for them to have some means of finding their position. If sailors can work out their latitude and longitude, then they know exactly where they are on earth.

LATITUDE

You know that the sun rises higher in the sky in the summer and lower in the winter. It is simple to measure the angle of the noon sun above the horizon. Once you know this angle, you can find the latitude. Angles have been calculated for every latitude for each day of the year. Books with this information are available. By looking up your angle's measurement in the tables, you can find your latitude.

"Shooting the sun" is the sailor's term for measuring the angle between the horizon and the sun with a sextant. It is very dangerous to look directly at the sun with the unaided eye, so a sextant has dark filters to make it safer. But incorrect use can still be dangerous. A sextant uses two mirrors to bring images of the sun and the horizon into line in a small telescope. When they line up, the angle between them can be read from a scale. Then the latitude can be looked up in the tables.

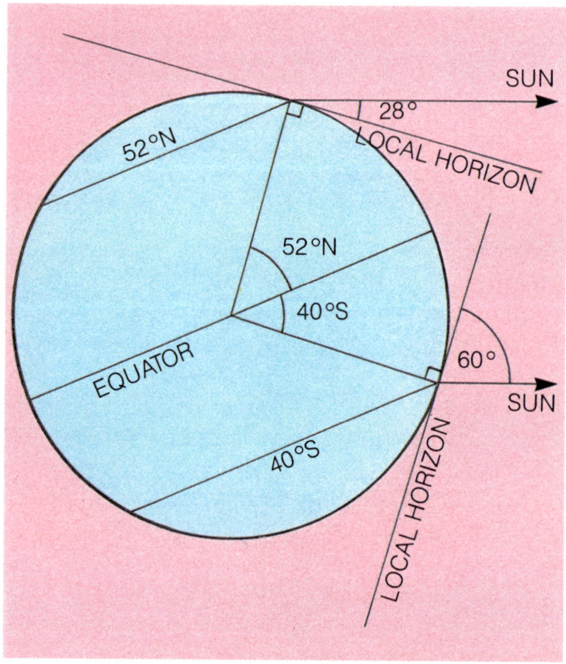

Above *A diagram showing how the angle of the noon sun can be calculated at latitude 52°N. On this day, the sun is 28° above the horizon. The drawing shows how this angle can be found at any latitude.*

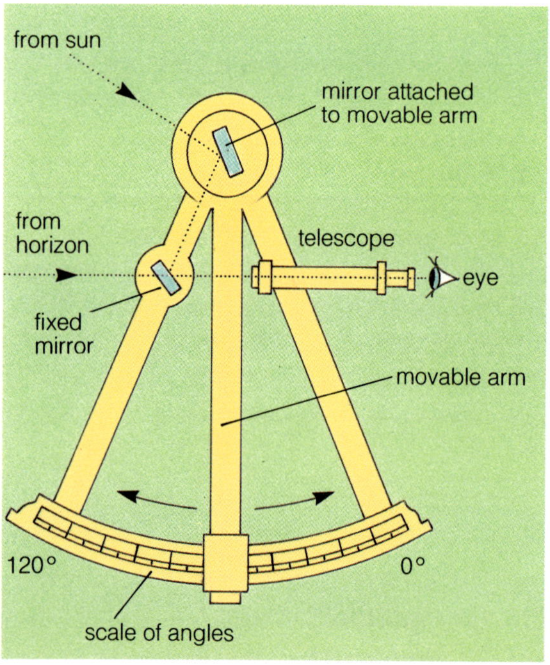

A diagram showing a sextant. This is often called an instrument of double reflection, because it has two mirrors.

LONGITUDE

You know that the Earth rotates through 15° every hour. This means that a particular time of day (for example, midday) "moves" around the Earth. Because of the direction of the Earth's spin, every hour westward of a given place is 15° of longitude (see page 20). Early navigators also knew this, but they could not make use of the information because they had no accurate way of keeping time.

To find your longitude, it is necessary to have a good means of telling time. You need to use a sextant to tell when it is noon by the sun wherever you are on Earth. You also need to know what time it is at Greenwich, which is 0° longitude. Using these two times, find the difference in hours between your location and Greenwich. Then multiply this time difference (in hours) by 15°. This gives your longitude in degrees. (See page 20.)

In 1760, John Harrison made the idea practical by inventing the first portable, reliable, and accurate ship's clock, called a chronometer. With such a clock, a ship's navigator always knows the exact time at Greenwich. This means that the crew can find their longitude, wherever they are.

By the end of the eighteenth century, sailors could plot their longitudinal position using John Harrison's invention, the chronometer.

TEST YOURSELF

1. What is "shooting the sun"?
2. How does a sextant work?
3. Why was the invention of the chronometer so important?

SPACESHIP EARTH

You have learned that the Earth is just one of nine planets moving around the sun, and that our sun is just one star among millions of other stars in orbit around our galaxy. You also know that our galaxy is just one among millions of other galaxies that make up the universe. Seen from this point of view, our planet seems very unimportant—just a tiny planet moving through the vastness of space.

However, our planet is very important to us—we depend on it completely. It has been said that the Earth is like a spaceship; that is a very good description. Every spaceship's crew must have clean air to breathe, protection from dangerous radiation (such as ultraviolet rays), supplies of food and water, a temperature-control system, and a good waste-recycling system. Like any spaceship, the Earth has to supply all the needs of its crew. If we (the crew) are to survive, we must take good care of our ship's "life-support" systems.

Of course, our planet's life-support systems are far more complicated than those of a spaceship, because the Earth contains so many different forms of life, all living together and depending on each other. But the principle is the same.

You have probably read or heard that some of these systems are being harmed: the ozone layer in the Earth's atmosphere, which protects us from ultraviolet rays; the tropical rain forests, which help to regulate world weather patterns; the rivers, seas, and lakes; and the air we breathe. Resources, such as coal, oil, and metals, are being used up at alarming rates and cannot be replaced.

These life-support systems have existed for millions of years, and the damage is being done by the ship's "crew"—humans. However, more and more people are becoming aware of the dangers created by human activity. For example, many countries have banned the use of CFCs (chlorofluorocarbons)—chemicals that damage the ozone layer. Much effort is being made to reduce the amount of air pollution, from power stations, factories,

Smog looms over the San Bernardino Mountains in California. This is air pollution belched out by power stations, factories, and cars.

It is important to protect our environment. One step that can be taken is to recycle waste, which usually contains precious resources. Here, household rubbish is being recycled at a special plant.

and cars. Governments and other groups are campaigning to get people to save energy, reduce and recycle waste, and use household chemicals that do not damage the environment.

These are just a few examples of how humans can save our planet's life-support systems from further destruction. Everyone can help in some way to keep our "spaceship" in good working order. After all, we really have no choice—there is nowhere else to go.

ACTIVITY

1 Find out about any plans in your area for helping preserve the environment. Some examples are waste recycling, antilitter campaigns, and energy conservation.
2 Find out all you can about the harm that is being done to our environment. See if you can discover what is being done, or suggested, to stop the damage. Page 44 will give you some ideas of areas to study.

Litter is not only ugly, but it can cause serious damage to the environment. However, it is easy to avoid littering.

TEST YOURSELF

1. Why is the description "Spaceship Earth" such a good one?
2. Name three of the Earth's life-support systems that are being harmed by human activity.

Glossary

Alien A person or creature from another world.

Asteroids Thousands of small, rocklike objects orbiting the sun between Mars and Jupiter. They range from 400 miles in diameter to less than 1 mile.

Astronomer A scientist who studies objects in space and the whole universe.

Comet A ball of ice, gas, and dust, moving in the solar system. Comets are sometimes seen to form long tails.

Crater A hole in the surface of a planet or moon, usually made by a meteorite.

Equator An imaginary circle running around the center of the Earth. It divides the Earth into the Northern and Southern hemispheres. Any line of longitude is at right angles to the equator. Any line of latitude is parallel.

Friction The force created by two objects or materials when rubbed together. The rubbing creates heat.

Galaxy A very large group of stars moving together in space.

Greenhouse effect This is caused when certain gases (e.g., carbon dioxide) form a layer above the Earth that traps heat and causes the atmosphere to heat up.

Greenwich Mean Time The local time at the 0° line of longitude which passes through Greenwich, England—a standard time for Britain and from which times throughout most of the world are calculated.

Horizon A distant point where the sky appears to meet the land.

Latitude The imaginary lines that run east to west right around the Earth. They are parallel to the equator.

Longitude The imaginary lines that run north to south around the Earth. They are at right angles to the equator.

Lunar Having to do with the moon.

Meteor A piece of rock that travels through space until it falls within the gravitational field of a star or planet.

Navigator The person on a ship (or any other form of transport) who works out the position and course to follow.

Nuclear fusion reaction The process of fusing atoms together to form new types of atom. During this process, large amounts of heat and light are released.

Orbit A curved path followed by one object (in space) around another, larger object.

Panorama The view all the way around the horizon.

Satellite An object in space that is in orbit around a larger body.

Solar Having to do with the sun.

Ultraviolet rays The rays from the sun that burn and brown the skin. They can cause serious harm to living things.

Unaided eye You eyesight without the use of a telescope, binoculars, or other optical instrument.

Books to Read

Bright Stars, Red Giants and White Dwarfs, Melvin Berger (Putnam, 1983)

Earth, Moon, and Stars, Cary I. Sneider (Lawrence Science, 1986)

The Galaxies: Cities of Stars, David J. Darling (Dillon, 1985)

The Moon, Heather Couper and Nigel Henbest (Watts, 1987)

Mysteries of the Planets, Franklyn M. Branley (Lodestar, 1988)

The Planets: Exploring the Solar System, Roy Gallant (Macmillan, 1990)

Planets, Stars, and Galaxies, Melvin Merger (Putnam, 1978)

The Space Telescope, Christopher Lampton (Watts, 1987)

Voyagers from Space: Meteors and Meteorites, Patricia Lauber (Harper/Collins, 1989)

Picture Acknowledgments

The author and publishers would like to thank the following for allowing illustrations to be reproduced in this book: Roger Browne *frontispiece,* 17; Mary Evans Picture Library 8; National Maritime Museum 16; PHOTRI *cover* (top right); TRH 22, 29, 30, 32, 35, 36; Wayland Picture Library 43; ZEFA *cover* (left), 6, 11, 18, 23, 24, 28 (right), 38, 44, 45. All inside artwork is by Jenny Hughes. Cover artwork is by Marilyn Clay.

Index

Air 6, 7, 27, 28, 44
Aldrin, Buzz 28
Andromeda Galaxy 38, 39
Apollo 11 28
Aristotle 8, 9
Armstrong, Neil 28
Asteroids 32
Astronomers 8, 34, 35, 36, 38, 40
Astronomy 9, 33
Atmosphere 34
Australia 10

Binoculars 28, 29, 34

Carbon dioxide 34
CFCs 44
Chronometer 43
Comets 32
Constellations 40-41
Copernicus 8, 9
Craters 27, 28, 29, 34

Equator 12, 14, 18
Equatorial dial 16, 17, 18

Galaxy 38-39, 40, 44
Galileo 8, 9
Garden sundials 18
Gnomon 16, 17, 18
Great Red Spot 34
Greenhouse effect 34
Greenwich 20, 21, 23, 43
  Mean Time 23

Harrison, John 43
Heat 24, 25, 36
Herschel, William 34
Horizontal sundial 18

Jupiter 32, 34

Latitude 12, 13, 14, 16, 18, 19, 20, 42
Light 10, 25, 36
Light-years 38, 40
Longitude 20, 21, 22, 23, 42, 43
Lunar eclipse 31
Lunar month 26, 27

Mars 32, 34
Mercury 32, 34
Meteors 28
Milky Way 38-39
Moon 8, 9, 26, 27, 28, 29, 30, 31, 32, 34, 38

Neptune 33, 34
Noon-dial 16
North Pole 6, 10, 13, 16, 20
Nuclear fusion reaction 25

Orbit 6, 14, 24, 31, 34, 44
Ozone layer 44

Phases of the Moon 26, 27
Planets 8, 9, 32-33, 34-35, 44
Pluto 33, 35, 36
Polar axis 16
Pollution 44
Proxima Centauri 36, 40
Ptolemy 9

Rain forests 44

Saturn 32
Seasons 10-13
Sextant 42
Solar eclipse 30-31
Solar system 32, 33, 34, 35, 36, 38
South Pole 10, 13, 16, 20
Space probes 34
Spin of the Earth 6, 8, 43
Stars 8, 30, 32, 36, 38, 40, 41, 44
Summer 10, 14, 15, 40, 42
Sun 6, 7, 8, 9, 10, 12, 14, 15, 16, 21, 24, 25, 26, 28, 30, 31, 32, 33, 34, 36-37, 38, 42, 43, 44
Sundials 16-19
Sunlight 6, 26, 28
Sunspots 36, 37

Telescope 28, 33, 34, 35
Time zones 22, 23
Tombaugh, Clyde 35

Ultraviolet rays 44
Universe 40, 41, 44
Uranus 33, 34

Venus 32, 33, 34

Water 6, 7, 27, 44
Winter 10, 14, 16, 42

First published in 1991 by
Wayland (Publishers) Ltd.